Wei / MDOT

STP 1193

Use of Waste Materials in Hot-Mix Asphalt

H. Fred Waller, editor

ASTM Publication Code Number (PCN)
04-011930-08

ASTM
1916 Race Street
Philadelphia, PA 19103

Library of Congress Cataloging-in-Publication Data

Use of waste materials in hot-mix asphalt / H. Fred Waller, editor.
(STP ; 1193)
Includes bibliographical references and index.
ISBN 0-8031-1881-3
1. Asphalt emulsion mixtures. 2. Waste products as road materials. 3. Pavements, Asphalt. I. Waller, H. Fred, 1926- II. American Society for Testing and Materials. III. Series: ASTM special technical publication ; 1193.
TE275.U84 1993
625.8'5--dc20 93-23992
CIP

Peer Review Policy

Each paper published in this volume was evaluated by three peer reviewers. The authors addressed all of the reviewers' comments to the satisfaction of both the technical editor(s) and the ASTM Committee on Publications.

To make technical information available as quickly as possible, the peer-reviewed papers in this publication were printed "camera-ready" as submitted by the authors.

The quality of the papers in this publication reflects not only the obvious efforts of the authors and the technical editor(s), but also the work of these peer reviewers. The ASTM Committee on Publications acknowledges with appreciation their dedication and contribution to time and effort on behalf of ASTM.

Printed in Baltimore, MD
June 1993

Foreword

This publication, *Use of Waste Materials in Hot-Mix Asphalt*, contains papers presented at the symposium on A Critical Look at the Use of Waste Materials in Hot-Mix Asphalt, held in Miami, FL on 8 Dec. 1992. The symposium was sponsored by ASTM Committee D-4 on Road and Paving Materials. H. Fred Waller of the Asphalt Institute in Raleigh, NC, presided as symposium chairman and is the editor of the resulting publication.

Contents

Petroleum Contaminated Soils

Polyethylene Waste

Papers Not Presented at Symposium (Publication Only)

Overview

The vast quantity of waste materials accumulating throughout North America is creating costly disposal problems. Some of these materials are by-products of industrial production processing, while others are waste materials from day to day usage by consumers. With governmental agencies becoming more environmentally conscious, it is a difficult and costly task to properly dispose of many of the materials in question, particularly where restrictive provisions prohibit their disposal in sanitary landfills. Some have been classified as hazardous wastes as authorities are concerned about possible contamination of ground water. These concerns lend impetus to exploration of alternate means of disposal; thus, the idea of this symposium was born. Its purpose is to examine the merits, both pro and con, of using several types of waste materials as one of the components in the production of hot-mix asphalt (HMA). The objectives of the symposium are fourfold, namely:

(1) To determine how waste materials should be processed and handled,
(2) To determine how various waste materials can be physically added to the asphalt mix,
(3) To determine the effects on mix properties and performance, and
(4) To determine the resultant cost increase in the finished product.

For the past several years, there have been limited studies to incorporate some of these waste materials into HMA. Materials involved to date include ground rubber tires, ground glass, asphalt shingles, contaminated sand/soils, incinerator ash, and various kinds of waste polymers. There are perhaps other waste materials that could be included in similar studies in the future. One governing criteria would be the quantity of material available for use. There must be a sufficient amount and a continuous supply in order for a specific material to be considered for use.

There are two primary factors that must be taken into account when the matter of incorporating waste materials into hot-mix asphalt are considered. One consideration is cost; there needs to be a balance between disposal of the waste material in the normal manner as compared to incorporation into the hot-mix asphalt. A second consideration is the effect on quality and performance of the HMA. It would be poor economics indeed to incorporate a waste material that substantially increases the cost of the HMA and at the same time shortens the service life or increases maintenance costs. Considerable additional research needs to be accomplished before we have satisfactory answers to these questions.

With the passage of the Intermodal Surface Transportation Efficiency Act (ISTEA) in Dec., 1991, a mandate was issued in Section 1038 that beginning in Jan. of 1994, ground automobile tires will be used in hot-mix asphalt. This act requires that at least 5% of Federal Aid hot-mix projects in 1994 must include some form of rubber tires. The percentage of Federal Aid projects requiring rubber will increase to 10% in 1995, 15% in 1996, and 20% in 1997, and each year thereafter. The ISTEA requirement stipulates that at least 20 lb of ground rubber per ton of asphalt mix shall be used. Each automobile tire produces approximately 12 pounds of available rubber for use in HMA.

Rubber tires, per se, probably offer the greatest potential for incorporation of a waste material. Reports indicate that there are about 3 billion tires currently in scrap heaps with about 240 million tires added each year. No longer can they be thrown into landfills or garbage dumps but must be transported to an approved disposal site. It has become normal practice for many agencies installing new tires on a vehicle to charge a fee to dispose of the old tires. Hopefully, an economic balance can be developed between using the scrap tires in HMA and the cost of their disposal through other approved means.

There are concerns, however, with respect to the unknown health effects on construction workers who may ingest fumes form asphalt-rubber mixes. Also, it is not known at this time whether pavements containing asphalt-rubber can be recycled to substantially the same degree as conventional HMA pavement; further, there is inadequate engineering data to predict performance of asphalt-rubber mixes. So, while there appears to be much potential for usage, there are many unanswered questions that must be resolved before a full endorsement by government and EPA authorities can be given.

Section 1038 of ISTEA requires the Secretary of the Department of Transportation (DOT) and the Administrator of the Environmental Protection Agency (EPA) to coordinate and conduct, in cooperation with the states, appropriate studies to provide answers to these questions. Such a report is due to Congress no later than 18 June, 1993. In view of these unanswered questions, The National Asphalt Pavement Association and the Asphalt Institute have made a formal request for a three year delay in the implementation of these ISTEA provisions. There may be an element of risk to move forward as planned until the industry concerns as stated have been properly addressed. Approval of this request should allow sufficient time for more comprehensive studies to be made.

Additional research should be initiated to provide indepth studies of the multitude of other available waste materials. We must have assurance that no waste material introduced into HMA will have a negative environmental impact, nor will it put construction workers at risk with respect to possible health effects. Leaders within the hot-mix industry have voiced strong concerns that our roads and pavements not become a dumping ground for waste materials simply to ease the burden of disposal or to comply with an unwarranted federal or state legislative requirement. There must be distinct, identifiable advantage(s) if any waste material is to be effectively used in HMA. We can ill afford to produce only "Trashphalt."

In an attempt to conserve both energy and materials, various kinds of recycling programs have been adopted by many public agencies. It is not uncommon to see collection bins scattered throughout major cities wherein wastes are separated into different categories. Some cities have established comprehensive recycling programs in an attempt to reclaim glass, aluminum cans, newspapers, and the like. New and innovative approaches must be developed from both an energy conservation and materials resources viewpoint. As an example, the New Jersey Department of Transportation (NJDOT) is offering HMA contractors a $1.00 per ton incentive on state paving jobs to include between 5 and 10% glass in either asphalt base or binder. NJDOT specifications allow up to 50% glass in these mixture types.

One must keep in mind the possibility of potential liability when new, nonstandard materials are used in the asphalt mix production process. Unexpected lawsuits may be a result. Further, the mix properties and specification requirements cannot be compromised to accommodate waste materials. The final answer with respect to mix performance and life cycle cost analysis will require a considerable time period before the economic feasibility and production procedures for using waste materials in HMA can be fully evaluated.

The reports contained in this Special Technical Publication (STP) are a major step in developing an understanding of some of the complexities associated with using waste materials in HMA. More research needs to be undertaken before specifications clerly defining the role of waste materials in the pavements construction industry can be developed. At this point, there appears to be signficant economic potential for effective use of mnay of our waste materials.

NOTE—As a result of the widespread interest shown in this symposium, ASTM is considering the formation of a new Committee entitled "Waste Materials." Subcommittees would be formed to deal with each individual type of waste material.

H. Fred Waller

Asphalt Institute, Raleigh, NC; symposium chairman.

Philosophy and Use of Waste Materials

Prithvi S. Kandhal[1]

WASTE MATERIALS IN HOT MIX ASPHALT - AN OVERVIEW

REFERENCE: Kandhal, P.S., "Waste Materials in Hot Mix Asphalt - An Overview," Use of Waste Materials in Hot-Mix Asphalt, ASTM STP 1193, H. Fred Waller, Ed., American Society for Testing and Materials, Philadelphia, 1993.

ABSTRACT: Numerous waste materials result from manufacturing operations, service industries, sewage treatment plants, households and mining. Legislation has been enacted by several states in recent years to either mandate the use of some waste materials or to examine the feasibility of such usage. The hot mix asphalt (HMA) industry has been pressured in recent years to incorporate a wide variety of waste materials into HMA pavements. This has raised the following legitimate concerns: (a) engineering concerns such as effect on the engineering properties (for example, strength and durability), impact on production, and future recyclability; (b) environmental concerns such as emissions, fumes, odor, leaching, and handling and processing procedures; and (c) economic concerns such as life cycle costs, salvage value, and lack of monetary incentives.

The waste materials can broadly be categorized as follows: (a)industrial wastes such as cellulose wastes, wood lignins, bottom ash and fly ash; (b) municipal/domestic wastes such as incinerator residue, scrap rubber, waste glass and roofing shingles; and (c) mining wastes such as coal mine refuse.

A general overview of preceding waste materials including the research work done in the past and their potential for use in HMA pavements is given in this paper.

KEY WORDS: hot mix asphalt, asphalt paving mixture, waste materials, crumb rubber, glass, roofing shingles

---

Numerous waste materials result from every aspect of society including manufacturing, service industries, sewage treatment plants, households and mining. The disposal of waste products is primarily done as follows:

(a) Landfills
(b) Incineration, and
(c) Recycling in other products

However, problems are being experienced because of the insufficient capacity of landfills, air pollution associated with

---

[1]Assistant Director, National Center for Asphalt Technology, Auburn University, Alabama

incinerators, and limited alternatives for recycling. Legislation has been enacted by several states in recent years to either mandate the use of some waste materials or to examine the feasibility of such usage. About 450 million megagrams (Mg) of hot mix asphalt (HMA) are produced in the United States at a cost of about $12 billion. The HMA industry has been pressured in recent years to incorporate a wide variety of waste materials into HMA pavements. This has raised the following legitimate concerns [1].

ENGINEERING CONCERNS

The following concerns must be addressed from the engineering viewpoint.

Properties of HMA

Since the waste material will replace and/or modify the properties of asphalt cement binder and/or HMA, it will affect the engineering properties (such as strength and durability). Therefore, the HMA containing the waste material must be reevaluated thoroughly and carefully both in the laboratory and the field.

Impact on Production

Some waste materials require a change in the HMA production equipment and/or processes and, therefore, the production is likely to be affected.

Future Recyclability

It is not known in many instances whether the HMA containing a waste material can be effectively recycled in the future without any problems. For example, HMA containing significant amounts of ground tire rubber has not been recycled as yet. Such recycling may pose air pollution problems.

Consistent Quality

The waste material may vary in quality (chemical or physical properties) and, thus, affect the engineering properties and durability of the hot mix asphalt.

ENVIRONMENTAL CONCERNS

The following concerns must be dealt with from an environmental viewpoint, especially when very tight emission and environmental controls are already in place on HMA production facilities.

Emissions/Fumes/Odor

It is quite likely that some waste materials will have emissions/fumes/odor problems because HMA is produced at high temperatures. These problems might also be faced when recycling HMA containing the waste material in the future.

Leaching

Component(s) of waste materials may be susceptible to leaching from HMA pavements and thus cause storm water and/or ground water pollution.

Handling and Processing Procedure

These procedures are not fully developed for all waste materials. Some waste materials could be hazardous and could pose health related risks to workers in the HMA industry.

ECONOMIC CONCERNS

From an economic viewpoint, the following concerns must also be addressed.

Price

Mandated use of waste materials is likely to raise the price of HMA. For example, virgin aggregates are generally available locally at lower costs compared to waste materials such as waste glass which must be hauled.

Life Cycle Costs

Life cycle costs need to be determined for HMA pavements containing waste materials. It is quite possible that the waste material may not affect the initial engineering properties of the HMA pavement but reduce its service life.

Disposal Costs

If the HMA containing the waste materials cannot be recycled, then its disposal costs need to be determined.

Salvage Value

Salvage values of HMA containing the waste materials are not known.

Lack of Incentive

At the present time there is a lack of monetary incentives to use the waste materials in HMA.

CATEGORIZATION OF WASTE MATERIALS

The waste materials can broadly be categorized as follows [2]. Some waste materials which have been used experimentally or routinely in HMA are also given for each category.

(a) Industrial Wastes

- Cellulose Wastes
- Wood Lignins
- Bottom Ash
- Fly Ash

(b) Municipal/Domestic Wastes

- Incinerator Residue
- Sewage Sludge
- Scrap Rubber
- Waste Glass
- Roofing Shingles

(c) Mining Waste

- Coal Mine Refuse

By-products from industrial processes, such as blast furnace slags, steel slags and sulphur were considered waste materials at one time. However, these products are now generally used on a routine basis and, therefore, will not be discussed. A detailed discussion of some selected waste materials used in HMA follows.

CELLULOSE WASTES

Cellulose wastes include agricultural wastes (crop residues, foresting wastes), manufacturing wastes (food processing, wood and paper industry), and urban refuse (municipal solid waste, manufacturing plant trash). A pyrolysis process has been used to convert cellulosic waste to a binder [3]. However, the resulting binder was not suitable for direct use as a substitute for asphalt cement because its rheological properties did not meet the performance criteria. The binder was also not compatible with asphalt cement. A pyrolysis-hydrogenation process, however, produced an oil which was suitable for use as an extender of asphalt cement in paving operations. Based on limited experiments the durability of a HMA wearing course mixture containing hydrogenated pyrolysis oil was not found significantly different from that containing the reference asphalt cement only. Extensive laboratory and field experiments are needed to evaluate the compatibility and performance of the cellulosic oils when used with asphalt cements from various sources.

WOOD LIGNINS

Wood lignin is a high volume waste produced in the manufacture of paper products. Attempts have been made to convert wood lignin to a highway binder material and use it either alone as a substitute for asphalt cement or as an extender for asphalt cement in HMA mixes [4]. Suitable lignin-asphalt binder formulations were prepared from lignins from both major manufacturing processes, that is, kraft and sulfite processes. None of the binder formulations was suitable alone as a substitute paving binder. However, a 30 percent replacement of asphalt cement appeared feasible with no significant effect on physical properties. Kraft lignin appeared insoluble in asphalt cement. Lignin-asphalt cement binders are stiffer and require slightly higher binder content than conventional HMA mixtures. Precoating of the coarse aggregate has been recommended to reduce the binder content in HMA mixtures. Based on limited laboratory experiments, Terrel et al. [4] concluded that HMA mixtures containing lignin-asphalt binders can be designed to match the structural strength of HMA mixtures containing conventional materials.

BOTTOM ASH

Bottom Ash is a waste material from coal burning power plants. It is the slag which builds up on the heat-absorbing surfaces of the furnace, and which subsequently falls to the bottom of the furnace and collected in an ash hopper. The bottom ash is categorized as dry bottom ash or wet bottom ash depending upon the boiler type used. The ash which is in solid state at the furnace bottom is called dry bottom ash. The ash which is in molten state when it falls in water is called either wet bottom ash or more commonly boiler slag [5].

Extensive laboratory studies were performed on representative samples of Indiana bottom ashes, to determine their physical, chemical and mechanical properties [6]. Depending upon the source, bottom ashes can have different physical and chemical properties. These studies showed that the bottom ashes are non-hazardous in nature and have minimal effects on ground water quality. However, the bottom ashes can be highly corrosive and, therefore, should not be placed very near to any metal structure.

Laboratory studies [7, 8] have also been conducted to evaluate the feasibility of using bottom ashes as a partial or full replacement of natural aggregates in HMA mixes and develop guidelines for their use. The properties of HMA containing bottom ash are dependent on ash content. Generally, as the ash content is increased, the optimum asphalt content is increased, the mixture density is decreased, and the air voids and voids in the mineral aggregate are increased. The stability of HMA mix decreases up to an ash content of 30 percent and then levels off. The mix containing bottom ash is susceptible to rutting. However, the mix is highly resistant to moisture induced damage (stripping). It was concluded from these studies that the properties of most dry and wet bottom ashes can meet the specifications for conventional aggregates.

Wet bottom ash (or boiler slag) can improve the skid resistance of HMA wearing courses. West Virginia obtained satisfactory performance from HMA wearing course mixes containing 50 percent boiler slag [9]. Lignite boiler slag has also been used to resurface residential streets in Texas. The HMA mix contained 75 percent boiler slag, 25 percent limestone screenings, and 6 to 7 percent asphalt content [10].

FLY ASH

Fly ash is the fine particulate matter which is precipitated from the stacks of pulverized coal-fired boilers at electrical power generating plants. It represents nearly 75 percent of all ash wastes generated in the United States. The quality of fly ash is affected by the type of coal used, the ash content of the coal, the degree to which the coal has been pulverized prior to combustion, and type of ash collectors utilized. These variables, therefore, impart a wide range of physical and chemical properties to the fly ash. It is composed of finely divided pieces of siliceous glass which range from 1 to 50 microns in diameter [11].

Fly ash has primarily been used as a mineral filler in HMA mixtures in several states such as Illinois, Michigan, Montana, North Dakota, and West Virginia [11]. Because the fly ash particles are extremely fine, it is quite likely to act as an extender of asphalt cement in HMA mixtures. Therefore, caution should be exercised when using fly ash in rut-resistant HMA mixtures for heavy duty pavements.

INCINERATOR RESIDUE

According to the estimates of the U.S. Environmental Protection Agency (EPA), about 23 million megagrams (Mg) of municipal solid waste (MSW) was incinerated in the United States in 1988 [12]. It is projected that about 41 million Mg and 50 million Mg of MSW will be incinerated in the years 1995 and 2000, respectively. Two ash products: bottom ash and fly ash, result from incinerating MSW. Bottom ash is the unburned and incombustible residue left on the boiler grates after incineration, and consists of large particles (0.1-100 mm) of slag, glass, rocks, metal, and other materials. Fly ash consists of burned or partially burned organic material particles which are usually 1-500 micron in size [5]. The heterogeneous character of these two incinerator residues causes a wide variation in composition from one plant to another.

The incinerator residue is not suitable for use in HMA mixtures unless it is processed further. The Franklin Institute Research Laboratory in Philadelphia developed a process of densifying (fusing) the unfused incinerator residue and converting it into useful construction materials. The incinerator residue is ground with a hammer mill, preheated at 688°C, and then melted and fused at 1093°C to form a column of semi-molten product which is later cooled and crushed to aggregate sizes [11]. Preliminary laboratory evaluation by the Pennsylvania Department of Transportation [13] indicated that the fused incinerator residue was not suitable for HMA mixtures because it did not comply with the specified gradation, particles were flat and elongated, and the HMA mixture did not meet design criteria for stability, air voids, or voids in the mineral aggregate. Subsequent improvements in the process [14, 15] produced a high quality aggregate material which performed well in a HMA surface course installation in Harrisburg, Pennsylvania [16].

Several installations using unfused incinerator residue as an aggregate in HMA mixtures were made during the 1974-1979 period in Texas, Pennsylvania, Washington, DC, and Massachusetts [2]. Details of these installations are given in the literature [17, 18, 19, 20, 21, 22]. The following recommendations have been made: (a) residues should be well-burned out (loss on ignition should be less than 10 percent), (b) HMA mixtures for base courses containing 50 percent natural aggregate and 50 percent incinerator residue hold the most promise, and (c) 2 percent lime should be added to minimize stripping problems.

TIRE SCRAP RUBBER

About 285 million tires are discarded every year in the United States. Of these, about 55 million are retreaded or reused (resold), and about 42 million are diverted to various alternative uses such as combustion for generating power and additive to HMA mixes. The remaining 188 million tires are added to stockpiles, landfills or illegal dumps. According to EPA estimates, 2 to 3 billion discarded tires are available at the present time [23]. Several states have enacted legislation to regulate the scrap tire problem. At the national level, section 1038 of the Intermodal Surface Transportation Efficiency Act (ISTEA) of 1991 specifically addresses the study and use of the scrap rubber by the highway industry.

Crumb rubber obtained from tires can either be ambient ground (grinding at room temperature or above) or cryogenically ground (grinding below embrittlement temperature, liquid nitrogen is often used). Ambient ground crumb rubber has a sponge-like surface. Due to very high surface area this rubber reacts with asphalt cement reasonably fast. Cryogenically ground rubber usually has undesirable particle morphology (structure). This process produces clean flat surfaces which, in turn, reduces the reaction rate with hot asphalt cement. Cryogenically ground rubber also gives lower elastic recovery compared to the ambient ground rubber [24].

The use of crumb rubber to modify asphalt cement has been developed over the past 25 years. Crumb rubber is primarily used in HMA mixes by two processes:

Wet Process (Asphalt-Rubber)

The wet process blends the crumb rubber with the asphalt cement prior to incorporating the binder into the project. The modified binder is commonly called "asphalt-rubber". Generally, 18-26 percent crumb rubber (16 mesh) by weight of asphalt cement is reacted with asphalt cement at 190 to 218°C for 1 to 2 hours. The blend is formulated at elevated temperatures to promote potential chemical and physical bonding of the two constituents. The first technology which applied the "wet process" is called the "McDonald Process". This process is also used for constructing stress absorbing membrane (SAM) and stress absorbing membrane interlayer (SAMI), and manufacturing crack sealers. SAM is a seal coat which uses asphalt-rubber as a binder. When SAM is placed as an interlayer it is called SAMI.

Dry Process (Rubber-Modified Mix)

This process mixes the crumb rubber with aggregate before incorporating the asphalt cement. About 3-5 percent of coarse rubber particles (1.6-6.4 mm) by weight of aggregate are generally used. The natural aggregate is usually gap-graded to accommodate the rubber particles as aggregate. The amount of crumb rubber used in the dry process can be 2-4 times that used in the wet process. The first application of the "dry process" in the United States is called the "PlusRide Process". It has been claimed that ice debonds easily from the pavement surface consisting of rubber-modified mixture because of higher than usual resiliency of the mix.

Details of Wet Process

The wet process (asphalt-rubber) will now be discussed in more detail. To produce an acceptable asphalt-rubber binder it is necessary to establish the digestion temperature and time for a specific combination of asphalt cement and crumb rubber. Viscosity of the blend is checked at different time intervals during the blending and digestion process. Viscosity of the blend increases with digestion time and then levels off. Achieving a reasonably constant viscosity indicates that the initial reaction is nearly complete and the binder is ready to use. This initial reaction is not well understood, but appears to be due to a chemical and physical exchange between the asphalt cement and rubber particles in which the rubber swells in volume causing an increase in viscosity. Continued mixing of asphalt cement and rubber after initial reaction can begin to reduce the viscosity of the blend as the rubber particles break down during mixing. However, breakdown of rubber particles is not rapid, and may require several hours at high temperatures.

If the asphalt-rubber blend is too viscous an extender oil is usually added. However, if less than 10 percent crumb rubber (by weight of asphalt cement) is used, extender oil is not needed.

The asphalt-rubber blend should be used as soon as possible after the initial reaction. Until used it should be recirculated continuously. Blends have been allowed to cool in storage tanks and reheated prior to use without difficulty.

When asphalt-rubber binder is used in HMA mixes the following mixing temperature ranges have been used:

Dense-graded HMA mixes: 163-190°C
Open-graded HMA mixes: 135-163°C

Stack emissions can be higher because of the elevated mixing temperatures.

HMA mixes containing asphalt-rubber binder should not be subjected to prolonged storage in a silo. Processing the mix through a surge silo is acceptable. Diesel fuel should not be used on truck beds as a release agent because it makes the mix stick more to the bed. Use of lime water, soap solution or silicone emulsion is recommended instead.

It is necessary to maintain the mix temperature within the desired range during placement of the mix. Static steel wheel rollers are generally used for compaction. Rubber-tired rollers tend to pick-up the mat. Vibratory rollers tend to tear and shove the mat. If the roadway has to be opened to the traffic right after compaction and the traffic has a tendency to pick-up the mat, it is recommended to apply 0.5-1 kg of concrete sand per square meter of the mat surface.

Recyclability of the HMA containing crumb rubber is a very important issue which must be addressed before using scrap tire crumb rubber on a large scale. There is no experience at the present time in recycling such mixes. It is quite likely that recycling these mixes in the future could pose air pollution problems during production. What will happen to the asphalt-rubber binder during recycling? How will the recycled mix be designed? If these concerns are not addressed it may not be possible to recycle the HMA mixes containing crumb rubber. In that case we will have to dispose the RAP (reclaimed asphalt pavement) which will be much worse than the problem of disposing tires.

In 1988, under a legislative mandate, the Florida Department of Transportation (FDOT) began a concerted effort to evaluate the potential use of tire scrap rubber in HMA pavements. FDOT commissioned the National Center for Asphalt Technology (NCAT) to prepare a state-of-the-art report on the use of ground tire rubber (GTR) in HMA and recommend a strategy to use GTR in Florida HMA mixes. This report [24] was published in August 1989. Since then, FDOT has constructed three demonstration projects. At the present time FDOT is implementing the following recommendations:

- Use GTR in friction courses only. Typically FDOT resorts to 75 mm cold milling (25 mm friction course + 50 mm structural course) which is recycled. If the friction course contains GTR, only 1/3 of the RAP will contain GTR.

- Dense-graded friction courses: Use 3-5 percent GTR by weight of asphalt cement. In the past, 18-20 percent GTR has been used as mentioned earlier. Use No. 80 mesh GTR. In the past, 16 mesh GTR has generally been used. Finer GTR is appropriate for the Florida friction course mixes which approach sand-asphalt mixtures. Moreover, the finer GTR size reacts with asphalt cement more easily, and thus a continuous blending of asphalt cement and GTR becomes possible.

- Open-graded friction courses: Use 5-10 percent GTR by weight of asphalt cement. Use No. 40 mesh GTR.

FDOT does a considerable amount of hot recycling every year. Since the RAP will contain low amounts of GTR, no problems are anticipated in recycling the HMA pavements in the future. It is estimated that state-wide use of GTR will increase the price of HMA mixes in Florida by 15 percent.

New Concepts

The "McDonald process" (wet process) and "PlusRide" (dry process) as discussed earlier are patented processes. However, the original patent on the "McDonald process" has expired in 1992. Recently some newer concepts which include the generic dry process, chunk rubber asphalt concrete and continuously blended asphalt rubber, have been introduced [23].

The generic dry process allows incorporation of crumb rubber in available "generic" aggregate gradation rather than gap-graded aggregate required in the PlusRide system. Experimental field applications of generic dry process have been made in New York, Florida, Iowa, Kansas, Oregon, and Illinois [23].

Chunk rubber asphalt concrete concept has been developed by the Cold Regions Research and Engineering Laboratory (CRREL) of the U.S. Army Corps of Engineers. This concept uses a large maximum crumb rubber size (4.75-12.5 mm), and an aggregate gradation to provide space for the rubber aggregate. However, this concept has been confined to laboratory only and no field installations have been made.

The continuous blending asphalt rubber concept has been used by Florida DOT (FDOT) as described earlier. It is a wet process which uses fine crumb rubber (180 μ or No. 80 mesh) to avoid reaction tanks, and to facilitate continuous blending. The performance of FDOT experimental sections is being monitored.

WASTE GLASS

When crushed waste glass is incorporated in HMA mix the resulting mixture is sometimes referred to as "glasphalt".

Several laboratory and field evaluations of glasphalt were conducted in the early 1970's in the U.S. and Canada. After no significant interest for a decade, the potential use of glasphalt is now being reassessed. Three states - Connecticut, Virginia and Florida - have recently conducted feasibility studies and issued research reports [25, 26, 27].

The Connecticut report gives an excellent review of literature on both laboratory and field evaluations of glasphalt since 1969:

- Glasphalt was successfully mixed and placed in at least 45 locations in the U.S. and Canada between 1969 and 1988. However, most glasphalt has been placed on city streets, driveways and parking lots, and not on high-volume, high-speed highways.

- Potential problems with glasphalt include: loss of adhesion between asphalt and glass; maintenance of an adequate level of skid resistance, especially with coarse particles; breakage of glass and subsequent ravelling under studded tires; lack of adequate and consistent supply of glass; and increased production costs (estimated at $5/Mg more than the conventional HMA mix in Connecticut).

- Glasphalt should be used only as a base course to alleviate potential skid resistance and surface ravelling problems.

- No more than a maximum of 9.5 mm glass should be used in glasphalt, with hydrated lime added to prevent stripping.

The Virginia report is based on laboratory evaluation and economic analysis of glasphalt. Two glass contents, 5 percent and 15 percent, and two asphalt contents were used in the laboratory evaluation of Virginia S-5 surface HMA mix (12.5 mm top size). A densely graded recycled glass with a maximum nominal size of 9.5 mm was used.

The following are significant conclusions:

- The use of glass tends to reduce the VMA and air voids in Marshall specimens; therefore, optimum asphalt content will also be reduced.
- Neither resilient modulus nor indirect tensile strengths are adversely affected by the addition of up to 15 percent glass.
- Although both wet strength and retained tensile strength ratios (TSR) were unaffected by the percentage of glass, some separation at the asphalt/glass interface was observed.
- A maximum of 15 percent crushed recycled glass should be allowed (100 percent passing a 9.5 mm sieve and a maximum of 6 percent passing a 75 μm sieve) in HMA mixes.
- There is little monetary incentive to use recycled glass at the present time because the cost of glass varies considerably.
- An experimental section should be laid prior to extensive use of waste glass.

The Florida Department of Transportation tested three HMA mixtures to determine the effects of crushed glass:

1. Control mix (9.5 mm nominal maximum size)
2. 15 percent Coarse glass mixture - same as the control mixture except that 15 percent of the screenings were replaced with coarse (9.5 mm - 2.06 mm nominal size) crushed glass.
3. 15 percent Fine glass mixture - same as the control mixture except that 15 percent of the screenings were replaced with fine (2.06 mm - 75 μm nominal size) crushed glass.

AC-30 asphalt cement with and without antistripping agent was used to prepare Marshall specimens which were also tested for tensile strength. The following conclusions were drawn based on the limited laboratory evaluation:

- Marshall stability values decreased by 15-20 percent and dry indirect tensile strength decreased by 20 percent when 15 percent of the screenings were replaced with either coarse or fine glass.
- Moisture conditioning of Marshall specimens caused a 15 percent and 50 percent decrease in tensile strength when coarse and fine glass, respectively, were incorporated

into the mixture. Retained tensile strength ratio (TSR) values indicated that the antistripping additive was ineffective in reducing the moisture damage.

- It is unlikely that the use of crushed glass in HMA mixtures will be economically feasible when suitable local materials are available at or near the HMA facility.

The City of New York places 360,000 Mg of HMA containing 10 percent waste glass every year. This amounts to an annual consumption of 36,000 Mg of glass.

ROOFING SHINGLES

Approximately 90 million roofing shingle squares are produced per year by 77 plants in the United States. About 1/3 of the shingles are used on new houses and the remaining 2/3 of the shingles are used for reroofing houses. When a house is reroofed, often an equivalent amount of old shingles is removed and discarded. Moreover, each roofing plant generates scrap materials and seconds that can range from 5 to 10 percent of the production capacity. The disposal of old shingles and the scrap material has created a difficult disposal problem [28].

It is estimated that roofing waste contains about 36 percent asphalt content, 22 percent hard rock granules (minus No. 10), 8 percent filler and smaller amounts of miscellaneous materials [29].

Roofing shingles have been used successfully in the HMA paving of the parking lots at Disney World, Florida. Shingles need to be shredded to at least 12.5 mm or smaller prior to introduction in the mix to ensure meltdown and uniform dispersion in the HMA mixture. According to cost estimates [28], the HMA cost can be reduced by $3.08 per megagram (Mg) by introducing only 5 percent organic shingles.

COAL MINE REFUSE

Anthracite Coal Refuse

The principal source of anthracite coal is in northeastern Pennsylvania. Anthracite coal mine refuse is composed of unprocessed mine refuse and processed coal breaker refuse. It is a gray slate-like material containing oxides of silicon, aluminum, iron, calcium, magnesium, sodium and potassium. More than 50 percent of average anthracite refuse is greater than 25.4 mm in size [11].

The Pennsylvania Department of Transportation (PennDOT) conducted a laboratory evaluation of unburnt (raw) anthracite refuse as an aggregate in HMA mixtures [30]. The aggregate particles were very absorptive, particularly in sizes larger than the 4.75mm sieve, and could not be coated completely with asphalt cement. Freeze-thaw failures also occurred in less than half the cycles of a control mix. Anthracite refuse must be incinerated before use in the HMA mix. Four HMA paving projects in Luzerne County of Pennsylvania were constructed with crushed incinerated anthracite refuse [31]. The performance in terms of skid resistance and wear was reported to be satisfactory.

Bituminous Coal Refuse

The principal sources of bituminous coal are located in Kentucky, West Virginia and Pennsylvania. Bituminous coal refuse is somewhat similar to anthracite refuse in appearance. The University of Kentucky evaluated the use of bituminous coal refuse in HMA mixtures in 1964 [32]. The mixtures containing 100 percent coal refuse were susceptible to moisture induced damage.

SUMMARY

Recycling of waste materials in highway construction should be encouraged. However, it is necessary to address the engineering concerns, environmental concerns and economic concerns mentioned in this paper before any large scale use of these materials. The use of waste materials should not be mandated. HMA containing a waste material should perform as well or better than conventional HMA. It should also be environmentally safe both for the first construction and future recyclability.

This paper discusses laboratory and/or field evaluation of several waste materials which have been used in HMA. It is recommended to construct demonstration projects to evaluate the performance of HMA containing waste materials. Waste materials should not be used as a standard practice until the performance data is available from the demonstration projects.

REFERENCES

[1] Warren, Jim, "The Use of Waste Materials in Hot Mix Asphalt." NAPA Special Report 152, June 25, 1991.

[2] Ormsby, W.C. and D.G. Fohs, "Use of Waste and By-Products in Highway Construction". Paper presented at the Annual Meeting of TRB, January 1991.

[3] Butte, W.A., E.M. Kohn and E.G. Scheibel, "Highway Binder Materials from Cellulosic and Related Wastes." FHWA Report No. FHWA/RD-130, December 1980.

[4] Terrel, R.L. et al., "Evaluation of Wood Lignin as a Substitute or Extender of Asphalt." FHWA Report No. FHWA/RD-80/125, October 1980.

[5] Ahmed, I., "Use of Waste Materials in Highway Construction." Joint Highway Research Report, Project C-36-50K, Purdue University, May 1991.

[6] Ke, T.C., "The Physical Durability and Electrical Resistivity of Indiana Bottom Ash." Research Report No. FHWA/IN/JHRP-9076, Purdue University, 1990.

[7] Majidzadeh, K., "Users Manual-Power Plant Bottom Ash in Black Base and Bituminous Surfacing: State-of-the-Art Report." FHWA Report No. FHWA-RD-77-148, September 1979.

[8] Majidzadeh, K., "Executive Summary-Plant Bottom Ash in Black Base and Bituminous Surfacing." FHWA Report No. FHWA-RD-79-72, September 1979.

[9] Moulton, L.K., Seals, R.K., and Anderson, D.A., "Bottom Ash: An Engineering Material." Journal of the Soil Mechanics and Foundations Division, American Society of Civil Engineers, Vol. 98, No. SM4, April 1972, pp. 311-325.

[10] "Lignite Slag Paves the Way." Industrial and Engineering Chemistry, Vol. 51, No. 7, July 1959, pp. 37A-38A.

[11] Miller, R.H. and Collins, R.J., "Waste Materials as Potential Replacements for Highway Aggregates." NCHRP Report 166, Transportation Research Board, 1976.

[12] "Characterization of Municipal Solid Waste in the United States: 1990 Update." Report No. EPA/530-SW-042, U.S. Environmental Protection Agency, 1990.

[13] Kandhal, P.S. and Wenger, M.E., "Incinerator Refuse as an Aggregate in Bituminous Concrete." Pennsylvania Department of Transportation, Bureau of Materials, Testing and Research, April 1971.

[14] Pindzola, D. and R.C. Chou, "Synthetic Aggregate for Incinerator Residue by a Continuous Fusion Process." FHWA Report No. FHWA/RD-74/23, 1974.

[15] Pindzola, D., "Large Scale Continuous Production of Fused Aggregate from Incinerator Residue." FHWA Report No. FHWA/RD-76/115, 1976.

[16] Snyder, R.R., "Evaluation of Fused Incinerator Residue as a Paving Material." FHWA Report No. FHWA-TS-88-229, 1980.

[17] Haynes, J., and Ledbetter, W.B., "Incinerator Residue in Bituminous Base Construction." Report FHWA/RD-76/12, Federal Highway Administration, 1976.

[18] Teague, D.J., and Ledbetter, W.B., "Three Year Results on the Performance of Incinerator Residue in Bituminous Base." Report FHWA/RD-78/144, Federal Highway Administration, 1978.

[19] Erdeley, J., and Ledbetter, W.B., "Field Performance of Littercrete (Incinerator Residue) in Bituminous Base." Report FHWA/RD-88/022, Federal Highway Administration, September, 1981.

[20] Pavlovich, R.D., Lentz, H.J., Ormsby, W.C., "Incinerator Residue as Aggregate for Hot-Mix Asphalt Base Course." Transportation Research Record 734, 1979, pp. 38-44.

[21] Griffith, J.L., "Summary Update of Research Projects with Incinerator Bottom Ash Residue." Commonwealth of Massachusetts, Executive Office of Environmental Affairs, Department of Environmental Management, Bureau of Solid Waste Disposal, Boston, MA, 1982.

[22] Turo, M.D., and Leonido, A.M., "Incinerator Residue as a Component of Bituminous Pavements." Massachusetts Department of Public Works Report R-37-0, 1984.

[23] Heitzman, M.S., "State of the Practice - Design and Construction of Asphalt Paving Materials with Crumb Rubber Modifier." Report FHWA-SA-92-022, Federal Highway Administration, May 1992.

[24] Roberts, F.L., P.S. Kandhal, E.R. Brown and R.L. Dunning, "Investigation and Evaluation of Ground Tire Rubber in Hot Mix Asphalt." NCAT Report 89-3, August 1989.

[25] Larsen, D.A., "Feasibility of Utilizing Waste Glass in Pavements." Connecticut Department of Transportation, Report No. 343-21-89-6, June 1989.

[26] Hughes, C.S., "Feasibility of Using Recycled Glass in Asphalt Mixes." Virginia Transportation Research Council, Report No. VTRC 90-R3, March 1990.

[27] Murphy, K.H., R.C. West, and G.C. Page, "Evaluation of Crushed Glass in Asphalt Paving Mixtures." Florida DOT Research Report No. FL/DOT/SMO/91-388, April 1991.

[28] Brock, J.D., and Shaw, D., "From Roofing Shingles to Roads." Technical Paper T-120, Astec Industries, Chattanooga, TN, 1989.

[29] Paulsen, G., Stroup-Gardiner, M., and Epps, J., "Roofing Waste in Asphalt Paving Mixtures." Center for Construction Materials Research, University of Nevada, Reno, 1988.

[30] Wenger, M.E. and Schmidt, G.H., "Anthracite Refuse as an Aggregate in Bituminous Concrete." Pennsylvania Department of Transportation, Research Project No. 70-8, June 1970.

[31] Mellott, D.B., "Anthracite Refuse in ID-2A Bituminous Concrete." Pennsylvania Department of Transportation, Research Project No. 70-8, July 23, 1970.

[32] Howard, G.G., "A Laboratory Investigation of the Properties of Coal-Bitumen Paving Mixtures." University of Kentucky, Engineering Research Bulletin No. 71, March 1964.

Stanley K. Ciesielski,[1] and Robert J. Collins[2]

CURRENT NATIONWIDE STATUS OF THE USE OF WASTE MATERIALS IN HOT MIX ASPHALT MIXTURES AND PAVEMENTS

REFERENCE: Ciesielski, S. K. and Collins, R. J., "**Current Nationwide Status of the Use of Waste Materials in Hot Mix Asphalt Mixtures and Pavements,**" Use of Waste Materials in Hot-Mix Asphalt, ASTM STP 1193, H. Fred Waller, Ed., American Society for Testing and Materials, Philadelphia, 1993.

ABSTRACT: Waste material management is becoming a very difficult challenge due to the lack of proper disposal facilities and processes as well as the environmental considerations associated with same. The objective of this paper was to research and report on the potential specific disposal of waste materials by usage in Hot Mix Asphalt Concrete (HMAC) mixtures. First, a search was made to identify all possible waste materials that may have potential for usage in an HMAC mixture. Secondly, all the fifty State Departments of Transportation and the District of Columbia were polled in order to establish their most current status on research and usage. Waste materials identified in this work included Reclaimed Asphalt Pavement (RAP), waste rubber and rubber tires, various slag and ash products and glass. The poll also identified those states or government agencies which are using waste materials as a matter of answering to existing or future legislative mandates.

KEYWORDS: Recycling, solid waste materials, rubber tires, slags, recycled asphalt pavement.

STATEMENT OF THE PROBLEM

The generation, handling and safe disposal of solid wastes has become a major concern in the United States. While the volume of wastes continues to grow, approval of

[1]Associate Professor, Civil Engineering Department, Villanova University, Villanova, PA 19085.

[2]President, R.J. Collins and Associates, P.O. Box 422, Springfield, PA 19064.

facilities for waste processing and proper disposal is becoming more difficult to obtain. At the same time, many existing disposal facilities are approaching capacity. Furthermore, environmental regulations have become increasingly more widespread and restrictive. As a consequence, the cost of waste handling and disposal has escalated significantly in recent years in many sections of the country.

Because of the expanding environmental awareness of the waste disposal problem, there is now more than ever before a definite trend toward recycling or reuse of a wide variety of solid waste materials in this country. Waste recycling in the 1990's has advanced to the recognition of the resource value in high volumes of formerly discarded materials such as old tires, paving rubble, combustion by-products, and other solid wastes. Because highways require huge volumes of construction materials, they have become a target of efforts to recycle or otherwise find uses for diverse waste materials.

The level of practice and knowledge regarding waste material usage in highway construction is not the same from one state to another. Because of growing pressures to recycle or reuse waste materials in the highway system, engineers and decision-makers at all levels need to be aware of the various types of waste materials, how or if they can be utilized in highway construction, experiences of others with such uses, and the technical considerations associated with their use. The specific focus of this paper is the mutually beneficial use of solid waste materials in the production and placement of Hot Mix Asphalt Concrete (HMAC) pavements in new highway construction or existing pavement rehabilitation using HMAC pavement overlays.

RESEARCH APPROACH

Literature Search

The research approach for this work involved a thorough review of published literature pertaining to the generation of various waste materials and by-products and the utilization of these materials in HMAC pavements. A number of different sources were consulted to develop the necessary information for this portion of the study. Included in the literature search were the following agencies and/or their publications.

- Transportation Research Information Service
- American Coal Ash Association
- American Society of Civil Engineers
- American Society for Testing and Materials
- Association of Asphalt Paving Technologists
- Federal Highway Administration

- o National Cooperative Highway Research Program
- o National Solid Waste Management Association
- o Transportation Research Board

In addition to the above sources, selected articles from the following periodicals were also reviewed.

- o Civil Engineering
- o Pit and Quarry
- o Public Works
- o Roads and Bridges
- o Rock Products
- o Solid Waste and Power
- o Waste Age

The following trade associations were contacted during the course of this investigation and provided technical literature or other related information:

- o American Coal Ash Association
- o American Concrete Pavement Association
- o National Asphalt Pavement Association
- o National Slag Association
- o Scrap Tire Management Council

State Department of Transportation Questionnaires

Questionnaires were sent to and were received from all 50 state departments of transportation (DOT) and the District of Columbia seeking information on the current state of practice relative to the usage of waste materials and by-products in HMAC pavement construction within each state. The questionnaire requested basic information from each state concerning:

- o Extent of current research on waste material usage
- o Acceptability of certain waste materials in HMAC pavements
- o Actual use of specific waste materials in HMAC pavement construction
- o Any waste materials considered unacceptable for HMAC pavement construction
- o Federal or state laws or mandates related to waste use.

INVENTORY AND CLASSIFICATION OF WASTE MATERIALS

Non-hazardous solid wastes and by-products which have been or have potential for use in HMAC can be classified according to source as:

- o Domestic
- o Industrial
- o Mineral

Domestic Wastes

Approximately 181 million metric tons of domestic wastes are generated annually in this country. Most of this waste is household or commercial trash and garbage, which is presently estimated at 168 million metric tons per year [1]. Currently, about 75 percent of trash or garbage is deposited in landfills, while 11 percent is recycled and 14 percent is burned [1]. The following domestic wastes have potential or actual usefulness as materials which are incorporated into HMAC highway pavements.

- o Incinerator Ash
- o Scrap Tires
- o Glass

Incinerator Ash--There are approximately 148 thermal reduction facilities in the United States with a capacity to burn at least 45 metric tons of solid waste per day. These facilities operate in 36 states and the District of Columbia. It is estimated that these facilities burn approximately 25.9 million metric tons per year of municipal solid waste (MSW) resulting in the generation of 7.8 million metric tons of incinerator ash or residue [2]. Approximately 90 percent of this ash is bottom ash and the remainder is fly ash. At present, most operating facilities combine the fly ash and bottom ash for disposal. Leachate analysis of selected incinerator ash grab samples indicates that most fly ash samples exceed regulatory limits for lead and cadmium, while the majority of combined ash samples do not exceed such limits [3]. Processed incinerator ash has been successfully used as a partial replacement for fine aggregate in HMAC mixes [4].

Scrap Tires--Approximately 235 million tires are discarded annually, comprising about 1.8 million metric tons of scrap rubber [5]. Over 80 percent of discarded tires are land-filled. Nearly 10 percent are recovered and used as tire derived fuel. About 2 percent of scrap tires are ground into crumb rubber and used in asphalt-rubber. It is estimated that as many as 2 to 3 billion scrap tires are stockpiled around the country. At least 30 states have legislation regulating the disposal of scrap tires [5].

Glass--Approximately 11.3 million metric tons of glass are included in the 168 million metric tons of household waste discarded annually. In 1988, a total of 1.4 million metric tons of glass was recycled [6]. The principal use of waste glass is as cullet for glass manufacturing. In highway construction, waste glass has potential as a partial replacement for fine aggregate in asphalt paving mixes [7].

<u>Industrial Wastes</u>

The annual generation of non-hazardous industrial wastes in the United States involves between 318 and 363 million metric tons of materials. Industrial wastes included in this study which have potential or have been used in HMAC pavements are as follows.

- o Coal ash By-Products
- o Iron and steel slags
- o Reclaimed asphalt pavement
- o Reclaimed concrete pavement
- o Foundry wastes

<u>Coal Ash By-Products</u>--Coal ash results from the burning of coal for power generation. The by-products resulting from coal combustion are fly ash, bottom ash, and boiler slag. There are approximately 420 coal-burning power plants located in 45 states. These plants generate nearly 65 million metric tons of coal ash annually, including 49 million metric tons of fly ash, 12.7 million metric tons of bottom ash, and 3.6 million metric tons of boiler slag, making coal ash one of the most plentiful mineral resources. Overall, only about 25 percent of all coal ash is used [8].

The principal highway uses of fly ash are in cement and concrete products, structural fills and embankments, road bases and subbases, grouting and asphalt fillers. Bottom ash and/or boiler slag have been used in asphalt paving [9].

<u>Iron and Steel Slags</u>--Blast furnace slag is the non-metallic by-product derived from producing iron in a blast furnace. Three basic types of blast furnace slag are produced: air cooled, granulated, and expanded. Air-cooled blast furnace slag is a fairly porous, lighter weight aggregate material. In 1989, a total of 14.1 million metric tons of blast furnace slag were produced, about 90 percent of which was air cooled. Blast furnace slag is produced in 13 states, primarily Pennsylvania, Ohio, Indiana, Illinois, and Michigan [7]. Air cooled blast furnace slag is an all-purpose construction aggregate commonly used in concrete, asphalt, road bases, as well as a fill material [9].

Steel slag is formed when lime flux reacts with iron ore, scrap metal, or other ingredients in a steel furnace. There are three basic types of steel furnaces (open hearth, basic oxygen, and electric arc), which produce three different types of steel slags. Approximately half of all currently operating steel furnaces are electric arc furnaces. However, many older slag banks contain open hearth slag. All steel slag is air cooled. During 1989, 7.2 million metric tons of steel slag were sold in the United States. There are steel slag processing locations in 26 states, although the largest quantities are produced in the leading blast furnace slag states. Steel slag has expansive tendencies unless properly aged with water. Steel slag has been used in asphalt paving, fill material, railroad ballast, and for snow and ice control [9].

Reclaimed Asphalt Pavement--It is estimated that approximately 90.7 million metric tons of asphalt paving material are currently being milled annually [8]. Much of this material is returned to producers' yards for future use in paving mixes. It is likely that only about 20 percent of all the milled asphalt material is recycled into hot mix asphalt the same year it is taken up [11]. Reclaimed Asphalt Pavement (RAP) can be recycled in hot mixes, cold mixes and in-place plant mixed [9].

Reclaimed Concrete Pavement--The American Concrete Pavement Association has indicated that approximately 322 kilometers of concrete pavement are being recycled each year Approximately 5,440 metric tons of reclaimed concrete can be reclaimed from every 1.6 kilometers of concrete pavement. This indicates that 2.6 million metric tons of reclaimed concrete are being recycled annually. Reclaimed concrete pavement (RCP) is useful as an unbound base course aggregate, in cement treated base, in asphalt paving, as an embankment base, and as rip rap [9].

Foundry Wastes--The principal types of foundry wastes include furnace dust, arc furnace dust, and sand reclaimer residue. The overall estimated quantity of foundry waste produced annually is believed to be from 9 to 14 million metric tons. There are approximately (2300) active foundry operations located in the United States, with Illinois, Wisconsin, Michigan, Ohio, and Pennsylvania being the states with the most foundries [12]. Foundry sand has been used as a fine aggregate in asphalt paving mixtures [9].

Mineral Wastes

Approximately 1.6 billion metric tons of mineral processing wastes are generated in the United States every year. In addition to these huge volumes, there are literally mountains of solid waste accumulations from past mining activities that are visible throughout many parts of the country. Mineral processing wastes used in HMAC can be classified as mill tailings.

Mill Tailings--Mill tailings are the finely graded waste products generated from ore concentration processes. Typically, tailings range from sand to silt-clay in particle sizing and are disposed of in slurry form by pumping into large ponds. Approximately 450 million metric tons per year of tailings are currently produced from milling operations. The largest amounts of tailings are generated from the concentration of copper, iron and taconite, lead, zinc, and uranium ores. Some sources of mill tailings have been utilized by state and local highway agencies in asphalt paving mixes [9].

Included in the category of mill tailings are quarry wastes, which include overburden and the fine screenings remaining from stone crushing or the washing of sand and gravel. It is estimated that approximately 68 million metric tons per year of quarry waste is being produced at stone quarries or sand and gravel pits.

TABLE 1--Classification and Annual Quantities of Solid Waste Materials and By-Products with Potential Use in Hot Mixed Asphalt Concrete Pavements

Waste Category	Description of Waste	Annual Quantity*	Total Annual Quantity*
Domestic	Household and Commercial Refuse	34	34
	- Glass (11.0)		
	- Plastics (13.0)		
	- Incinerator Ash (7.8)		
	- Scrap Tires (2.2)		
Industrial	Coal Ash	65	206
	Blast Furnace Slag	15	
	Steel Mill Slag	7	
	Non Ferrous Slags	9	
	Reclaimed Asphalt Pavement	91	
	Reclaimed Concrete Pavement	3	
	Foundry Wastes	9	
	Roofing Shingle Waste	7	
Mineral	Waste Rock	925	1,395
	Mill Tailings	470	
		Total	1,635

*Millions of Metric Tons

CURRENT STATE DOT RESEARCH

In conjunction with the many HMAC uses for different waste materials and by-products discussed, there are a number of research activities related to waste material or by-product utilization in HMAC pavements. In the State DOT questionnaire, distributed in June, 1991, state materials engineers indicated the extent of research performed in their respective state on uses of waste materials or by-products in HMAC and potential acceptability of such uses. Based on the questionnaire responses from all 50 states, a total of 41 states have performed or plan to perform some research into using one or more waste materials or by-products in HMAC pavements.

Research performed by state departments of transportation on the most commonly investigated waste material and/or by-products involved at least 10 different HMAC related applications. Table 2 provides a list of these applications, presented alphabetically according to code letters. Table 3 is a summary of research activities performed by state departments of transportation for these 10 different waste materials, with applications studied for each waste or by-product indicated by the appropriate code letter denoted in Table 2.

TABLE 2--List of Possible Uses for Waste Materials and By-Products Evaluated by State Departments of Research in HMAC and Pavements

Code	Description of Use or Application
ACM	Asphalt cement modifier
AGG	Aggregate in asphalt
AR	Asphalt-rubber
CS	Chip Seal
JCS	Joint and crack sealant
MF	Mineral filler in asphalt
OS	Overlay sealant
REC	Recycled pavement as aggregate in HMAC
SAM	Stress absorbing membrane
SND	Sand substitute

According to Table 3, the wastes or by-products used in HMAC most frequently researched were:

- Reclaimed asphalt pavement (RAP) (32 states)
- Scrap tires (34 states)
- Coal ash (10 states)
- Blast furnace slag (11 states)
- Recycled glass (7 states)

Of the states that researched RAP, 32 evaluated its use as an aggregate in asphalt pavement recycling (hot or cold).

TABLE 3A.--Summary of State Departments of Transportation Research Activities on Uses of Waste Materials and By-Products in HMAC

States	Fly Ash	Bottom Ash	Blast Furnace Slag	Steel Making Slag	RAP
1. Alabama*	...	...	...	...	...
2. Alaska	...	...	...	...	...
3. Arizona	...	...	...	...	REC
4. Arkansas	...	AGG	...	...	REC
5. California	...	...	AGG	AGG	REC
6. Colorado	...	...	...	...	REC
7. Connecticut	...	...	...	...	...
8. Delaware*	...	...	...	...	...
9. Florida	...	...	...	...	REC
10. Georgia	...	...	...	...	REC
11. Hawaii*	...	...	...	...	...
12. Idaho*	...	...	...	...	...
13. Illinois	...	...	...	...	...
14. Indiana	AR	...	AGG	AGG	REC
15. Iowa	...	...	...	...	REC
16. Kansas	MF	...	AGG	AGG	REC
17. Kentucky*	...	...	...	...	...
18. Louisiana	MF	...	AGG	AGG, CS	REC
19. Maine	...	...	...	...	REC
20. Maryland	...	...	...	...	REC
21. Massachusetts	...	...	...	...	REC
22. Michigan	...	...	...	...	REC
23. Minnesota	...	...	...	...	...
24. Mississippi	...	...	...	...	REC
25. Missouri	...	AGG	AGG	AGG	REC

TABLE 3A.--(Continued)

States	Fly Ash	Bottom Ash	Blast Furnace	Steel Making	RAP
26. Montana*	...	...	...	...	...
27. Nebraska	MF	...	...	...	REC
28. Nevada	...	...	...	...	REC
29. New Hampshire	...	...	...	...	REC
30. New Jersey	MF	...	...	...	...
31. New Mexico*	...	...	...	...	...
32. New York	MF	...	...	...	REC
33. North Carolina	...	...	...	...	REC
34. North Dakota*	...	...	...	...	...
35. Ohio	...	...	...	...	REC
36. Oklahoma	...	...	...	...	...
37. Oregon	...	...	...	...	REC
38. Pennsylvania	...	...	AGG	AGG	REC
39. Rhode Island	...	...	...	...	REC
40. South Carolina	...	...	AGG	AGG	REC
41. South Dakota*	...	...	...	...	...
42. Tennessee	...	...	AGG	...	REC
43. Texas	...	AGG	AGG	...	...
44. Utah	...	...	...	...	REC
45. Vermont	...	...	...	...	...
46. Virginia	...	...	AGG	...	REC
47. Washington	...	...	...	...	REC
48. West Virginia	...	...	AGG	AGG	REC
49. Wisconsin	...	...	...	...	...
50. Wyoming	...	AGG	...	...	REC
District of Columbia	...	...	...	...	...

*Has not performed any recent research.

TABLE 3B.--Summary of State Department of Transportation Research Activities on Uses of Waste Materials and By-Products in HMAC

States	Scrap Tires	Glass	Foundry Wastes	Mill Tailings
1. Alabama*	...	...	...	...
2. Alaska	AR	...	...	...
3. Arizona	AR	...	...	...
4. Arkansas	AR	...	...	...
5. California	AR, CS	AGG	...	...
6. Colorado	SAM	...	...	...
7. Connecticut	AR	AGG	...	...
8. Delaware*	...	...	...	...
9. Florida	AR	...	...	...
10. Georgia	JCS	...	...	...
11. Hawaii*	...	...	...	...
12. Idaho*	...	...	...	...
13. Illinois	...	...	SND	...
14. Indiana	...	...	AGG	...
15. Iowa	AR	AGG	...	...
16. Kansas	AR	...	...	AGG
17. Kentucky*	...	...	...	...
18. Louisiana	...	...	...	...
19. Maine	AR, CS	...	...	...
20. Maryland	AR	...	...	...
21. Massachusetts	AR	...	...	...
22. Michigan	AR	...	...	...
23. Minnesota	AR	...	...	...
24. Mississippi	AR	...	...	...
25. Missouri	AR	...	...	AGG

TABLE 3B.--(Continued)

States	Scrap Tires	Glass	Foundry Wastes	Mill Tailings
26. Montana*	...	...	...	...
27. Nebraska	AR	...	...	...
28. Nevada	CS	...	...	AGG
29. New Hampshire	SAM	...	...	...
30. New Jersey	...	...	...	...
31. New Mexico*	...	...	...	...
32. New York	AR, JCS	AGG	...	AGG
33. North Carolina	AR	...	...	...
34. North Dakota*	...	...	...	...
35. Ohio	ACM	...	...	...
36. Oklahoma	AR	...	...	AGG, CS
37. Oregon	AR	...	...	...
38. Pennsylvania	AR	AGG	AGG	...
39. Rhode Island	AR	...	...	...
40. South Carolina	AR, CS	...	AGG	...
41. South Dakota*	...	...	...	...
42. Tennessee	JCS	...	...	...
43. Texas	AR	...	...	...
44. Utah	...	...	...	...
45. Vermont	...	AGG	...	...
46. Virginia	AR	AGG	...	...
47. Washington	AR	...	...	...
48. West Virginia	...	...	...	...
49. Wisconsin	ACM	...	AGG	...
50. Wyoming	AR	...	...	...
District of Columbia	ACM	...	...	...

*Has not performed any recent research.

Scrap tires were investigated for use in asphalt-rubber mixes (wet or dry) by 27 different states. Four states investigated the use of ground tires in chip seals. Other applications for scrap tires included joint sealant and stress absorbing membrane.

Air-cooled blast furnace slag was primarily researched as an aggregate in asphalt mixes (11 states). Steel making slag has been investigated in HMAC mainly as an aggregate in asphalt paving. Coal bottom ash has been evaluated as an aggregate in hot-mix asphalt. Five states have researched the use of fly ash as a mineral filler in asphalt paving. Recycled glass was primarily researched for its possible use as a fine aggregate in hot-mix asphalt paving (7 states).

CURRENT STATE DOT USE OF WASTE MATERIALS

The state DOT questionnaire also requested information in the actual use of waste materials and by-products in HMAC. The information obtained from the state DOT questionnaire supplemented a survey of waste materials used in highway construction, that was published in mid-1991 by Purdue University [13]. The information presented in the Purdue report was also obtained by means of state DOT questionnaires. Waste usage reported in HMAC in this paper includes reported uses of waste materials from the Purdue study.

As in the research activities section, uses of waste materials and/or by-products by the state departments of transportation in HMAC have involved different highway related HMAC applications. Table 2 provides a list of these applications presented alphabetically according to code letters. Table 4 is a summary of the various waste materials and or by-products that have been used at one time or another in HMAC by state transportation agencies. According to Table 4, a total of 18 different waste materials or by-products have actually been used in HMAC at some time by four or more state transportation agencies. Table 5 is a list of all waste materials or by-products used by the State DOT agencies as reported in the questionnaire. As seen from Table 5, 44 states have indicated use of reclaimed asphalt pavement (RAP). A total of 38 states have used scrap tires in rubber-asphalt mixes. In all, a total of 16 different waste materials or by-products have been utilized by at least one state in HMAC according to the questionnaire responses.

TABLE 4A.--Summary of State Department of Transportation Utilization of Waste Materials and By-Products Used in HMAC

States	RAP	RCP	Scrap Tires	Glass
1. Alabama	REC	AGG	...	...
2. Alaska	...	...	AR	...
3. Arizona	...	...	AR	...
4. Arkansas	REC	...	...	...
5. California	REC	...	AR	AGG
6. Colorado	REC	...	...	...
7. Connecticut	REC	AGG	AR	...
8. Delaware	REC	...	JCS	...
9. Florida	REC	...	AR	...
10. Georgia	REC	...	JCS	...
11. Hawaii	REC	...	...	...
12. Idaho	...	...	AR	...
13. Illinois	REC	...	JCS	...
14. Indiana	REC	...	AR, JCS	...
15. Iowa	REC	AGG	AR	AGG
16. Kansas	REC	...	AR	...
17. Kentucky	REC	...	...	...
18. Louisiana	REC	...	...	...
19. Maine	REC	...	JCS, CS	...
20. Maryland	REC	...	...	...
21. Massachusetts	REC	...	ACM, AGG	...
22. Michigan	REC	REC	AR	...
23. Minnesota	...	...	AR	...
24. Mississippi	REC	...	...	...
25. Missouri	REC	...	AR	...

TABLE 4A.--(Continued)

States	RAP	RCP	Scrap Tires	Glass
26. Montana	REC	REC	AGG	...
27. Nebraska	REC	...	AR	...
28. Nevada	REC	...	CS	...
29. New Hampshire	REC	...	SAM	...
30. New Jersey	...	...	AGG, JCS	AGG
31. New Mexico	REC	...	AR, SAM	...
32. New York	REC	...	AGG	AGG
33. North Carolina	REC	...	AR	...
34. North Dakota	REC	AGG	AR	...
35. Ohio	REC	...	AR	...
36. Oklahoma	REC	...	AR, JCS	...
37. Oregon	REC	...	AR	...
38. Pennsylvania	REC	...	AR	...
39. Rhode Island	REC	...	AGG	...
40. South Carolina	REC	...	CS	...
41. South Dakota	REC	REC	...	...
42. Tennessee	REC	...	JCS	...
43. Texas	REC	...	CS, AR	...
44. Utah	REC	...	...	...
45. Vermont	...	...	...	AGG
46. Virginia	REC	...	AR	AGG
47. Washington	REC	...	AR, AGG	...
48. West Virginia	REC	...	...	...
49. Wisconsin	...	...	AR	...
50. Wyoming	REC	...	AR, JCS	...
District of Columbia	...	...	ACM	...

TABLE 4B.--Summary of State Departments of Transportation Utilization of Waste Materials and By-Products Used in HMAC

States	Fly Ash	Bottom Ash	Blast Furnace Slag	Steel Making Slag	Mill Tailings
1. Alabama	...	...	AGG	AGG	AGG
2. Alaska	...	...	...	...	...
3. Arizona	...	...	...	...	...
4. Arkansas	...	...	...	...	...
5. California	...	...	AGG	AGG	...
6. Colorado	...	...	...	...	...
7. Connecticut	...	...	...	...	...
8. Delaware	...	...	...	...	...
9. Florida	...	...	...	...	...
10. Georgia	...	...	AGG	...	...
11. Hawaii	...	...	...	...	...
12. Idaho	...	...	...	...	...
13. Illinois	...	...	...	...	...
14. Indiana	...	...	AGG	AGG	...
15. Iowa	...	...	...	...	...
16. Kansas	MF	...	...	...	AGG
17. Kentucky	...	...	AGG	AGG	...
18. Louisiana	MF	...	CS	AGG	...
19. Maine	...	...	...	...	...
20. Maryland	...	...	...	...	...
21. Massachusetts	...	...	...	...	...
22. Michigan	...	...	...	...	...
23. Minnesota	...	...	REC	...	AGG
24. Mississippi	...	AGG	...	...	...
25. Missouri	...	AGG	AGG	AGG	AGG

TABLE 4B.--(Continued)

States	Ash	Bottom Ash	Blast Furnace	Steel Making	Mill Tailings
26. Montana	...	...	...	...	...
27. Nebraska	MF	...	...	...	...
28. Nevada	...	...	...	...	AGG
29. New Hampshire	...	...	...	...	...
30. New Jersey	MF	...	...	...	...
31. New Mexico	...	...	...	...	...
32. New York	MF	SND	...	...	AGG
33. North Carolina	...	...	...	...	...
34. North Dakota	...	...	...	...	...
35. Ohio	...	...	AGG	...	...
36. Oklahoma	...	...	...	...	AGG
37. Oregon	...	...	...	...	...
38. Pennsylvania	...	...	AGG	AGG	...
39. Rhode Island	...	...	...	...	...
40. South Carolina	...	...	AGG	AGG	...
41. South Dakota	...	...	...	...	...
42. Tennessee	...	...	...	...	...
43. Texas	...	AGG	AGG	...	...
44. Utah	...	...	...	...	...
45. Vermont	...	...	...	...	...
46. Virginia	...	...	...	...	...
47. Washington	...	...	...	...	...
48. West Virginia	...	...	AGG	AGG	...
49. Wisconsin	...	...	...	...	...
50. Wyoming	...	...	...	...	...
District of Columbia	...	...	...	...	...

TABLE 5--List of State DOT Usage of Waste Materials as Indicated from Questionnaire Responses in HMAC

Waste Material	Number of States Using
Reclaimed Asphalt Pavement	44
Scrap Tires	38
Iron and Steel Slags	13
Reclaimed Concrete Pavement	7
Mining Wastes	7
Waste Glass	6
Coal Fly Ash	5
Coal Bottom Ash	4
Plastic Waste	3
Kiln Dusts	2
Incinerator Ash	2
Roofing Shingle Waste	2
Broken Concrete	1
Foundry Waste	1
Quarry Waste	1
Non Ferrous Slags	1

In addition to the end uses indicated in Table 4, it must be pointed out that some of these end uses are considered by the states to be experimental. These states have indicated that one or more of the waste materials being used in their state are being used for testing or demonstration purposes only.

Scrap tires have been used in asphalt-rubber paving mixes either as part of an asphalt-rubber binder (wet process) or as a fine aggregate substitute in gap graded mixes (dry process), in about 60% of the states. The dry process has been used in 10% of these states. In 40% the states in which scrap tires have been used in asphalt, their use is still considered experimental by the state department of transportation. In 25% of the states, scrap rubber use is not considered successful by the state department of transportation, either because of poor performance or because its use is not economical.

SPECIFICATIONS FOR WASTE MATERIALS AND BY-PRODUCTS

Table 6 provides a breakdown of the different specifications that have been prepared and/or are being used by each state. At least 18 different waste materials or by-products are included in one or more state specifications. Table 6 shows the different major wastes or by-products that are specified and the number of states in which these materials are specified. This table indicates all end uses that may be specified for each waste material or by product. The wastes or by-products most frequently included in state specifications are RAP, fly ash, scrap tires, blast furnace slag, RCP, and steel slag.

TABLE 6--Most Frequently Used State Specifications Including Waste Materials or By-Products in HMAC

Description of Material Specification	Number of States Specifying
o Reclaimed asphalt pavement as aggregate in new or recycled asphalt mixes	48 states
o Granulated tire rubber in asphalt-rubber paving mixtures or in stress-absorbing membrane interlayers	22 states
o Fly ash as mineral filler in asphalt	20 states
o Air-cooled blast furnace slag as an aggregate in asphalt mixes	16 states
o Steel slag as an aggregate in asphalt wearing surface mixes	14 states
o Reclaimed concrete pavement as an aggregate in new asphalt pavement	12 states
o Granulated tire rubber in asphalt-rubber seal coats	11 states
o Crushed glass as a fine aggregate in asphalt paving mixes	5 states

FINDINGS AND CONCLUSIONS

1. Reclaimed asphalt pavement (RAP) is the most frequently specified and used by-product in HMAC.

2. Although nearly three dozen states have used rubber from scrap tires in some form of asphalt paving, more than one-third of those states still consider their use of scrap tires to be experimental. There are at least a half dozen states that do not intend to continue using scrap tires because they consider their use either unsuccessful or uneconomical.

3. Blast furnace slag is an all-purpose construction aggregate that has been used with considerable success in asphalt paving.

4. Mining wastes, in particular some sources of waste rock and mill tailings, are suitable as aggregate materials for base courses and for asphalt paving, depending on their chemistry and availability.

REFERENCES

[1] National Solid Waste Management Association, "Landfill Capacity in the Year 2000," Washington, D.C. (1989).
[2] Kiser, J.L., "A Comprehensive Report on the Status of Municipal Solid Waste Combustion," Waste Age (November, 1991).
[3] Repa, E., "The Confusion and Questions About Ash," Waste Age (September, 1987).
[4] Turo, M.D. and Leonido, A.M, "Incinerator Residue, A Component of Bituminous Pavements," Massachusetts Department of Public Works, Draft Final Report, Wellesley Hills, Massachusetts (December, 1984) 24 pp.
[5] National Solid Waste Management Association, "Waste Product Profile: Tires," Washington, D.C. (1990).
[6] National Solid Waste Management Association, "Recycling in the States. 1990 Review," Washington, D.C. (1990).
[7] Malisch, W.P., Day, D.E. and Wixson, B.G., "Use of Domestic Waste Glass as Aggregate in Bituminous Concrete," Highway Research Board, Record No 307, Washington, D.C. (1970) pp. 1-10.
[8] American Coal Ash Association, "1989 Coal Combustion By-Product Production and Consumption," Washington, D.C.
[9] Ciesielski, S.K. and Collins, R.J., "Recycling and Use of Waste Materials and By-Products in Highway Construction," National Cooperative Highway Research Program, Synthesis of Highway Practice - In Press, Transportation Research Board, Washington, D.C. (1993).
[10] Owens, J.F., "Slag-Iron and Steel," U.S. Bureau of Mines, Mineral Yearbook, Washington, D.C. (1989).
[11] Krissoff, M.R., Asphalt Recycling and Reclaiming Association, Annapolis, Maryland, Private Communication.
[12] Mosher, G., American Foundrymens Society, Chicago, Illinois. Private Communication.
[13] Ahmed, I., "Use of Waste Materials in Highway Construction," Purdue University, Joint Highway Research Project, Report No. FHWA/IN/JHRP-91/3, West Lafayette, Indiana (May 1991), ppg. 114.

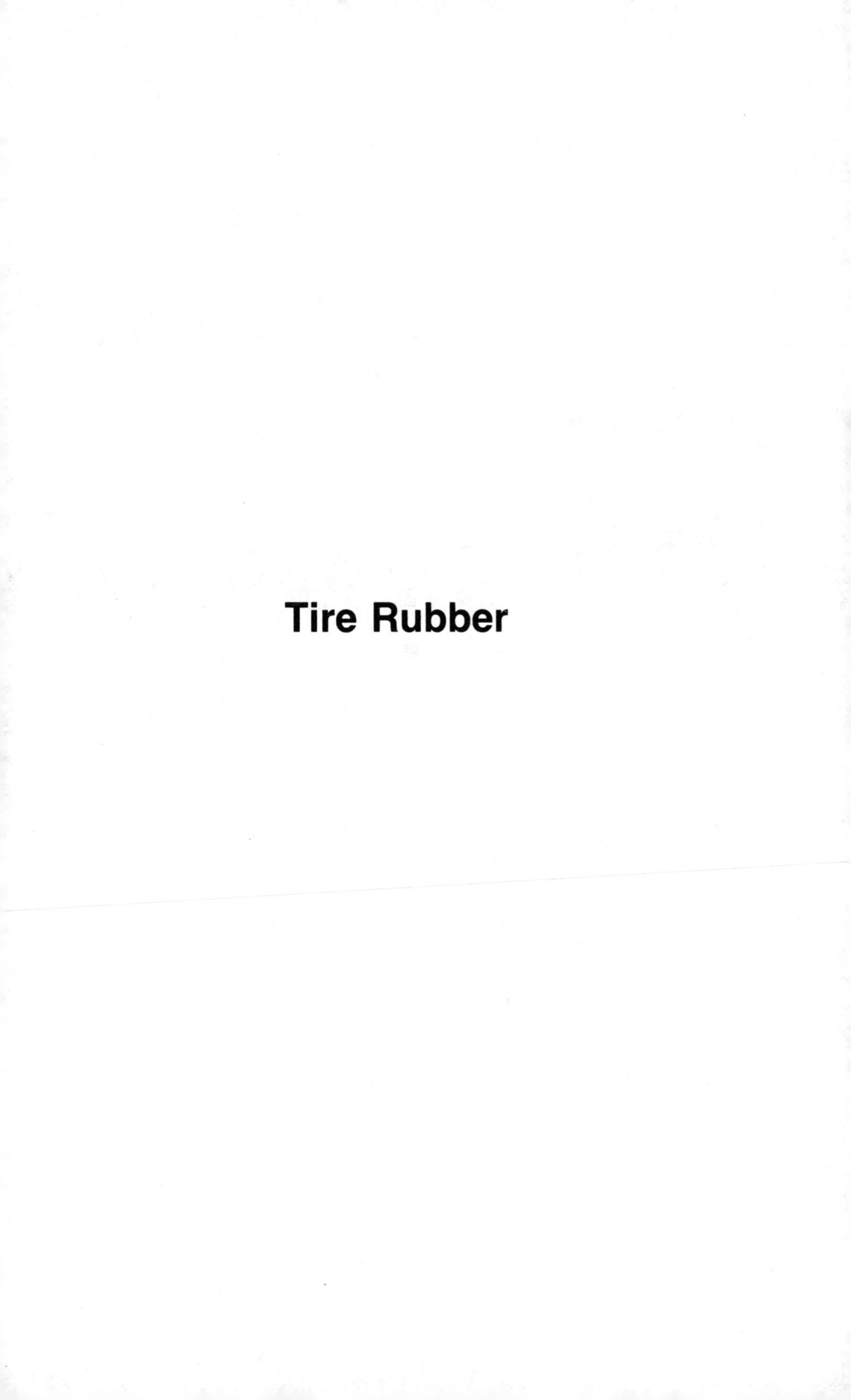

Tire Rubber

Scott Shuler[1] and Cindy Estakhri[2]

RECYCLED TIRE RUBBER AS AN ASPHALT MODIFIER

REFERENCE: Shuler, S. and Estakhri, C., **"Recycled Tire Rubber as an Asphalt Modifier,"** Use of Waste Materials in Hot-Mix Asphalt, ASTM STP 1193, H. Fred Waller, Ed., American Society for Testing and Materials, Philadelphia, 1993.

ABSTRACT: A combination of ground tire rubber and asphalt cement was evaluated in the laboratory and in two full-scale experimental pavements. Tire rubber was added to asphalt in quantities from 18 to 26 percent by weight including whole tire and tread rubber produced from both ambient and cryogenic grinding processes. Rubber gradation was varied to provide graded and one-sized distributions of rubber particles in the resulting blends of asphalt-rubber.

Asphalt-rubber placed in the experimental pavements and comparable blends prepared in the laboratory are evaluated by laboratory tests. Four principal tests are presented for evaluation of asphalt-rubber behavior including force-ductility, double-ball softening point, rotational viscosity, and size exclusion chromatography. Results of laboratory testing indicates properties of field prepared asphalt-rubber can be duplicated in laboratory prepared blends. Condition survey results from three full-scale test pavements provide information on the most effective combinations of asphalt and tire rubber for interlayer construction.

A rotational viscometer was developed which simultaneously blends the rubber and asphalt and monitors changes in consistency. The variation in consistency with increasing rubber content is presented as a possible means of monitoring rubber concentration during construction.

Chemical analysis of the asphalt and rubber blends by gel permeation chromotography indicates that some chemical modification occurs to the asphalt as a function of blending temperature and time.

A method is presented which describes an extraction process for asphalt and rubber. Extraction of rubber from asphalt-rubber blends after various digestion periods indicates that a loss of integrity of solid rubber occurs as a function of digestion time, temperature and rubber type and size. Periods up to 24 hours resulted in a loss of solid rubber by weight of the blend of up to 34 percent due to digestion with asphalt.

KEY WORDS: Asphalt-rubber, recycled tire rubber, ground tire rubber, force ductility test, stress absorbing membrane interlayer (SAMI), chip seals.

[1] Executive Vice President, Colorado Asphalt Producers Ass'n, Denver, CO.
[2] Research Engineer, Texas Transportation Institute, College Station, TX.

INTRODUCTION

Ground tire rubber has been used commercially as an additive in various types of asphalt pavement construction since the early 1970's. This use began by blending ground tire rubber with asphalt at elevated temperatures and applying the resulting modified asphalt as a binder for chip seals and interlayer systems. Later, this ground tire modified asphalt or *asphalt-rubber* was used in asphalt concrete applications and later still, the rubber was added to the asphalt concrete as a dry elastic 'aggregate'. This paper documents a field and laboratory experiment using the so-called *asphalt-rubber* products, or wet process binders, in interlayer construction.

Asphalt-rubber has been defined as a blend of ground tire rubber and asphalt cement at 18 to 26 percent rubber by total weight of the blend [1]. The blend is formulated at elevated temperature to promote chemical and physical bonding of the two components. Various petroleum distillates are sometimes added to the blend to reduce viscosity and promote workability.

An asphalt-rubber chip seal sandwiched between an existing cracked asphalt concrete pavement and new asphalt concrete overlay is called an asphalt-rubber "interlayer" [2]. Observations of field installations of over two hundred separate pavement sections containing asphalt-rubber have indicated that asphalt-rubber bound materials reduce the occurrence of reflection cracking when used as interlayers in certain applications [1].

Many types of asphalt-rubber formulations are possible due to a wide assortment of constituents available. Evidence suggests certain asphalt-rubber blends may produce undesirable results in the laboratory [3]. Although some data are available regarding performance of asphalt-rubber in the laboratory [3, 4, 5, 6, 7] better correlation between laboratory data and field performance is needed.

Objective

The purpose of this research was to design and construct two field test pavements containing asphalt-rubber interlayers to determine the objective value of using asphalt-rubber for reducing reflective cracking. Pavement condition surveys conducted prior to interlayer construction provide data regarding initial pavement condition. These data establish a datum which will allow future comparison of field performance for the various types of interlayer blends placed at each test road.

Laboratory tests were performed on blends of asphalt-rubber prepared in the field as well as blends prepared in the laboratory. These data form the basis for future correlations between laboratory properties and field performance.

TEST ROADS

Two field test pavements were constructed as part of this research. One test pavement was constructed in the east and westbound travel lanes on Interstate 10 east of El Paso, Texas for approximately nine miles between FM 34 and the McNary interchange. This pavement will be referred to as the "El Paso Test Road".

The second test pavement was constructed in the northbound travel lane of Interstate Highway 45 from the Leon-Freestone County Line north to the U.S. 84 overpass, a distance of approximately eighteen miles. This pavement will be referred to as the "Buffalo Test Road".

MATERIALS

El Paso Test Road

Asphalt cement used in the preparation of asphalt-rubber binders was obtained from the Chevron refinery in El Paso, Texas. This asphalt meets the Texas State Department of Highways and Public Transportation (SDHPT) specification [12] requirements for AC-10 as shown in Table 1.

TABLE 1 -- Asphalt properties.

Property	El Paso	Buffalo	Specification Min	Specification Max
Viscosity, 60C, Pa-s (P)	105 (1048)	87 (868)	80 (800)	120 (1200)
Viscosity, 135C, cSt	290	280	190	
Penetration, std, dmm	92	150	85	
Flash Point, C (F)	316 (600+)	n/a	232 (450)	
Specific Gravity, 25C	1.010	1.017	n/a	

Three sources of rubber were used to produce asphalt-rubber binders investigated at the El Paso Road. These rubber materials were obtained from the suppliers shown in Table 2.

TABLE 2 -- Rubber sources and types.

Rubber	Source	Designation	Description
A El Paso	Genstar Chandler, AZ	C104 - El Paso	Whole Tire, Vulcanized, Ambient Grind
B El Paso	Atlos Los Angeles, CA	TPO - 44	Tread Tire, Vulcanized, Ambient Grind
C El Paso	Midwest Elastomers Wapokonetta, OH	n/a	Whole Tire, Vulcanized, Cryogenic Grind
D Buffalo	Genstar Chandler, AZ	C106 - Buffalo	Whole Tire, Vulcanized, Ambient Grind

Gradation of the rubber appears in Figure 1.

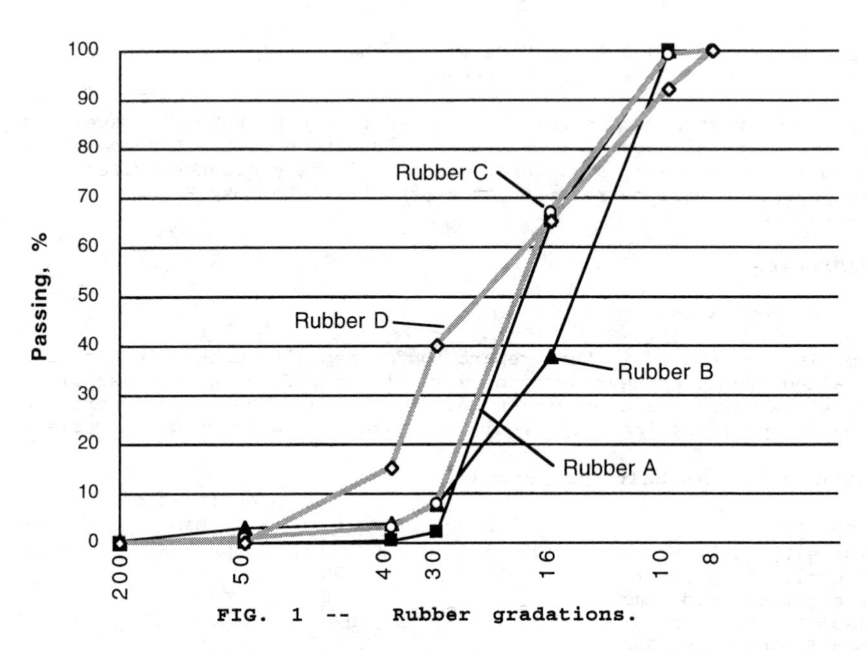

FIG. 1 -- Rubber gradations.

Further characterization of each rubber type following ASTM procedure D297 [11] provides data relating to physical and chemical properties as shown in Table 3.

TABLE 3 -- Rubber properties.

Property	A	B	C	D
Specific Gravity	1.165	1.153	1.150	1.160
Total Extract, w%	15.45	19.47	24.50	15.41
Ash, w%	5.71	3.49	2.41	5.68
Free Carbon, w%	29.21	30.75	31.31	29.00
Sulfur, w%	1.17	1.02	1.10	1.15
Rubber Polymer, w% RHC[1]	30	20	0	30
SBR, w% RHC	60	80	55	60
Polybutadiene, w% RHC	10	0	45	10
RHC, v%	60.92	55.89	50.76	61.02

Dolomite mineral aggregates used in construction of the interlayer were obtained from the Esperanza Pit, Esperanza, Texas. Interlayer aggregates were precoated with approximately one percent Chevron AC-20 and stockpiled prior to application.

Particle size gradations of interlayer aggregates appear in Figure 2.

1 RHC, Rubber Hydrocarbon Content

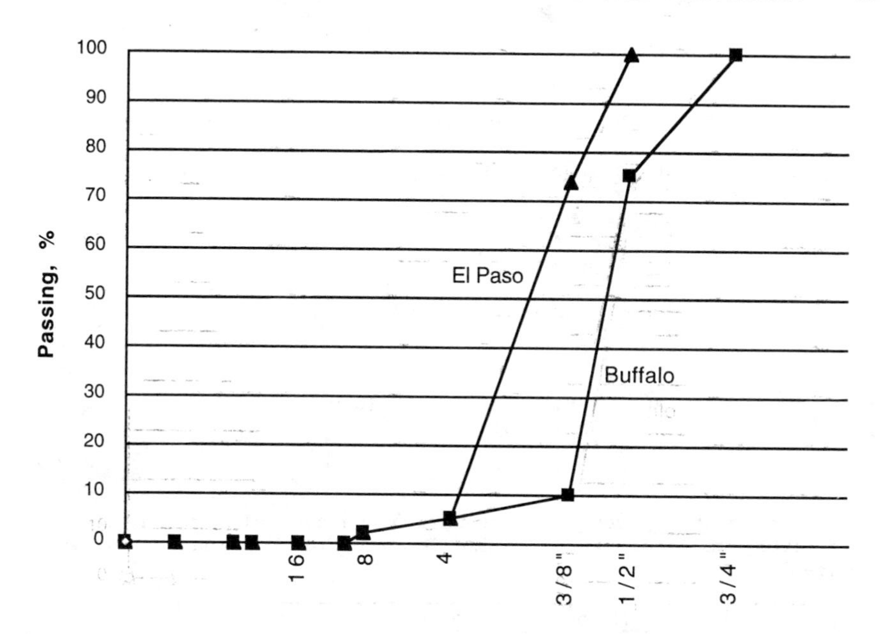

FIG. 2 -- Interlayer aggregates.

Buffalo Test Road

Asphalt used for asphalt-rubber blending was an AC-10 asphalt cement supplied by Texas Fuel and Asphalt, Corpus Christi, Texas. Asphalts met the Texas SDHPT specification requirements for AC-10 viscosity graded materials as shown in Table 1. A flux oil, Sundex 790, from Sun Oil Corporation, Houston, Texas, was blended with the AC-10 asphalt prior to blending with rubber.

One rubber source was used to produce the asphalt-rubber placed on the Buffalo Test Road. This material is described as Rubber D Designation C106 in Table 2. This rubber has the same chemical properties as Rubber Type A Designation C104 used at the El Paso Test Road. However, particle size gradation differs. Sieve analysis of the rubber is shown in Figure 1.

Limestone mineral aggregates used for construction of interlayer and asphalt concrete were obtained from the Yelberton Pit near Mexia, Texas. Interlayer aggregates were precoated with approximately 0.50 percent AC-20 immediately prior to application.

Particle size gradations of aggregates are shown in Figure 2.

EXPERIMENT DESIGN

El Paso Test Road - Field Responses

Independent variables studied in the field experiment are as follows:

I. Rubber Type, Ri
  A. Type A
  B. Type B
  C. Type C
II. Rubber Concentration, %, Cj
  A. 22
  B. 24
  C. 26
III. Application Rate, l/m2 (gsy), Ak
  A. 1.59 (0.35)
  B. 1.81 (0.40)
  C. 2.04 (0.45)

Nine experimental pavement sections and one control section (no interlayer) were constructed using the material combinations shown in Figure 3.

Application Rate, lsm(gsy)	Rubber Concentration, %		
	22	24	26
1.59 (0.35)	C Section 2	B Section 9	A Section 8
1.81 (0.40)	B Section 4	A Section 1	C Section 6
2.04 (0.45)	A Section 5	C Section 7	B Section 3

Rubber Type

FIG. 3 -- El Paso test sections.

El Paso Test Road - Laboratory Experiment

The laboratory experiment evaluates materials prepared in the field and materials prepared in the laboratory.

Field PreparedAsphalt-Rubber

Concentration, %	Rubber Type: A	B	C
26	- - -	- - -	- - -
24	- - -	- - -	- - -
22	- - -	- - -	- - -

FIG. 4 -- Combinations of field prepared asphalt-rubber.

LaboratoryPreparedAsphalt-Rubber

Three digestion conditions were produced in the laboratory as shown below:

Digestion Level	Temperature, C (F)	Blending Conditions Time, min.
Low	163 (325)	30
Moderate	177 (350)	60
High	191 (375)	180

These digestion conditions were varied from low to high to provide a range from which simulation of field digestion could be approximated. The basis for this lab variation was an effort to provide asphalt-rubber lab mixes with properties of field prepared mixes.

Digestion	Rubber Type: A 22	A 24	A 26	B 22	B 24	B 26	C 22	C 24	C 26
High	- - -	- - -	- - -	- - -	- - -	- - -	- - -	- - -	- - -
Med	- - -	- - -	- - -	- - -	- - -	- - -	- - -	- - -	- - -
Low	- - -	- - -	- - -	- - -	- - -	- - -	- - -	- - -	- - -

FIG. 5 -- Combinations of laboratory prepared asphalt-rubber.

Buffalo Test Road - Field Experiment

Levels of the independent variables are as follows:

I. Concentration of Rubber, Ci
   A. 18
   B. 22
II. Digestion, Dj
   A. Low
   B. High

In this experiment, rubber type and application rate are held constant. The resulting four treatments are replicated providing eight experimental sections. Four additional test sections were included as control sections resulting in a total of 12 test sections. Two sections were constructed using a conventional asphalt cement as the interlayer binder and the other two sections contain no interlayer.

Buffalo Test Road - Laboratory Responses

Levels of the independent variables are as follows:

I. Concentration of Rubber, Ci
   A. 18
   B. 22
II. Digestion, Dj
   A. Low
   B. Medium
   C. High

This experiment was designed to evaluate laboratory responses of laboratory mixed asphalt-rubber materials. These replicates will allow future comparison of field performance within a given treatment such that variability can be judged between treatment types. In this study, it was desired to see whether laboratory responses differed significantly for replicate materials fabricated in supposedly the same manner.

SITE SELECTION

Location of both field test roads was accomplished in cooperation with the Texas SDHPT. A list of sites was obtained from highway districts planning asphalt-rubber interlayer construction and from this list potential test sites were selected. Criteria used to judge the adequacy of sites are listed in order of importance below:

1. Willingness of district and contractors to participate in experiment.
2. Size of project.
3. Time until next planned rehabilitation.
4. Pavement substructure uniformity.
5. Overlay thickness and uniformity.
6. Distress uniformity.

A contract had been awarded on the project which would become the El Paso Test Road when initial contact with the El Paso Highway District was made. Since significant changes in the original contract were required to accomodate the planned experiments, it was crucial that a cooperative spirit exist between highway department, contractor, and research personnel. Planning the Buffalo Test Road began before there was a contract between the highway department and a contractor. Therefore, requirements of test section construction were included in job specifications and subject to competitive bidding.

A full distributor of asphalt-rubber was desired for use in application of each test section for both test roads. This was desirable for reasons listed below:

1. A more representative blend of asphalt-rubber could be expected compared with partial loads,
2. Test section length of approximately one lane-mile resulted from approximately 15920 l (4200 gal) distributor loads. These lengths provided transitions before and after the 457.5 m (1500 ft) of photologs contained in each test section. This further enhanced the potential for representative materials placed over photologs.
3. Production rate was not appreciably slowed. This enhanced the desired cooperative spirit between contractor and research personnel.

Project size was an important factor for both test roads since it was desired to place test sections in lanes having consistent traffic volumes and loads. Both projects were of sufficient length to accomodate approximately nine lane-miles for the El Paso Test Road and over ten lane miles for the Buffalo Test Road.

El Paso Test Road

The El Paso Test Road is part of Texas Project FR-10-1(168)079 located on Interstate Highway 10 (IH-10) in Hudspeth County, approximately 129 km (80 mi) east of El Paso between the McNary interchange and FM 34 as shown on Figure 6. Test sections are each approximately 1.45 km (0.90 mi) in length in the travel lanes as shown in Figure 7.

Original pavement structure for eastbound lanes was U. S. Highway 80 consisting of a 6.1 m (20 ft) wide portland cement concrete pavement constructed in 1932. Conversion of the original highway to the interstate system in 1963 added westbound lanes consisting of 15 cm (6 in) dense graded asphalt concrete over 15 cm (6 in) cement treated base and 15 cm (6 in) cement treated subgrade. An overlay of original portland cement concrete pavement in 1963 consisted of 15 cm (6 in) dense graded asphalt concrete in which 7.6 cm (3 inch) by 15 cm (6 inch) Number 10 welded wire fabric was embedded in the lower 3.8 cm (1-1/2 in).

Distress consisted of slight to severe transverse cracking at random intervals, and combinations of longitudial and alligator cracking distributed throughout.

Traffic on the El Paso Test Road consisted of a total traffic volume of 7900 average daily traffic (ADT) in 1983. Truck volume was approximately 25 percent of this value with five axle semi-trucks accounting for approximately 60 percent of all trucks.

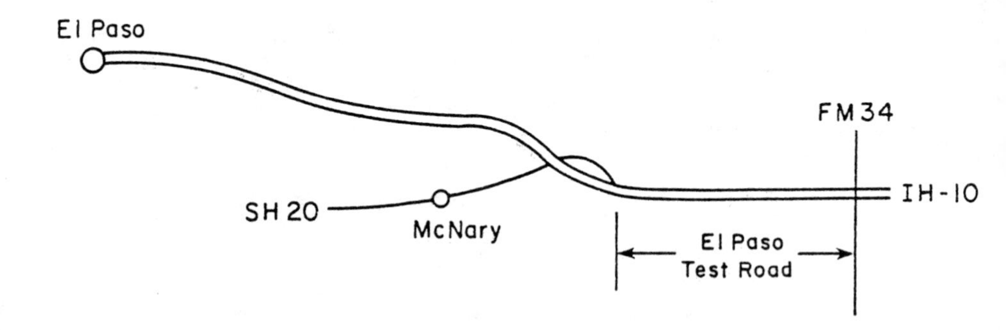

Figure 6 - El Paso test road location

Subgrade soils on the El Paso Test Road are poorly graded sands and gravels, some containing plastic fines, classified by the Unified Soil Classification System as GP-GC and SP-SC for gravels and sands, respectively.

Buffalo Test Road

Buffalo Test Road State project designation is FRI-45-2(68)180 located on Interstate Highway 45 (IH-45) in Freestone County, from the Leon county line to US 84 as shown in Figure 8. Test sections are each approximately 1.29 km (0.80 mi) in length in the northbound travel lane as shown in Figure 9.

The Buffalo Test Road is constructed on 20.3 cm (8 in) of continuously reinforced concrete pavement over 10 cm (4 in) of asphalt treated basecourse and 15 cm (6 in) lime treated subgrade. The original pavement structure was constructed in 1971.

Distress consisted of typical hairline random transverse cracks at 0.9 cm (3 ft) to 1.8 cm (6 ft) intervals, and infrequent punchouts.

Traffic on the Buffalo Test Road was measured by Texas SDHPT in 1983 at approximately 15,000 ADT. The total volume of trucks is approximately 20 percent, Volume by individual truck type has not been measured in this area and is therefore, not available.

Subgrade soil types along the Buffalo Test Road alignment were obtained from recently recorded Soil Conservation Service logs [16]. Classification of subgrade soils by the Unified System are as low plasticity clays and silty clays, ML-CL, along much of the alignment with some clays bordering on high plasticity.

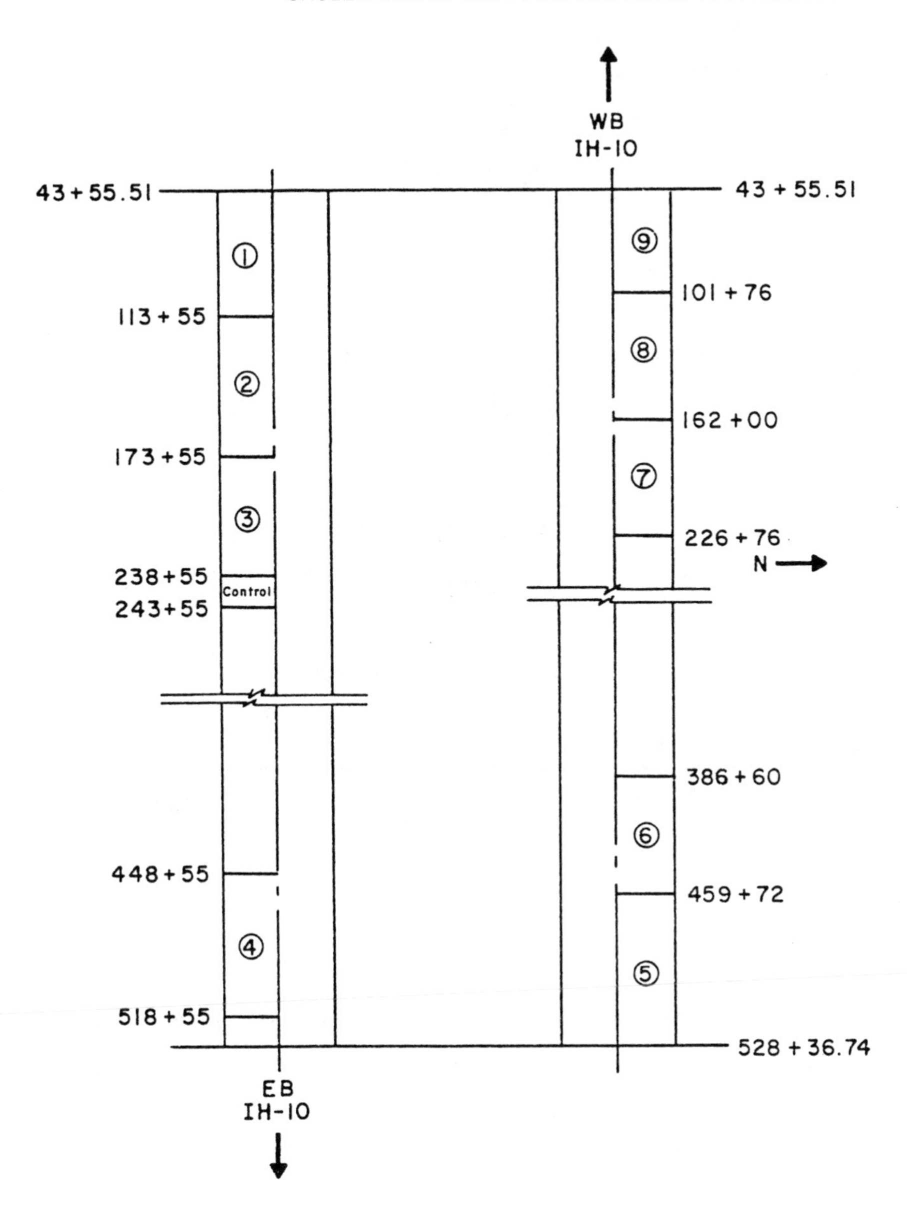

Figure 7 - El Paso test sections

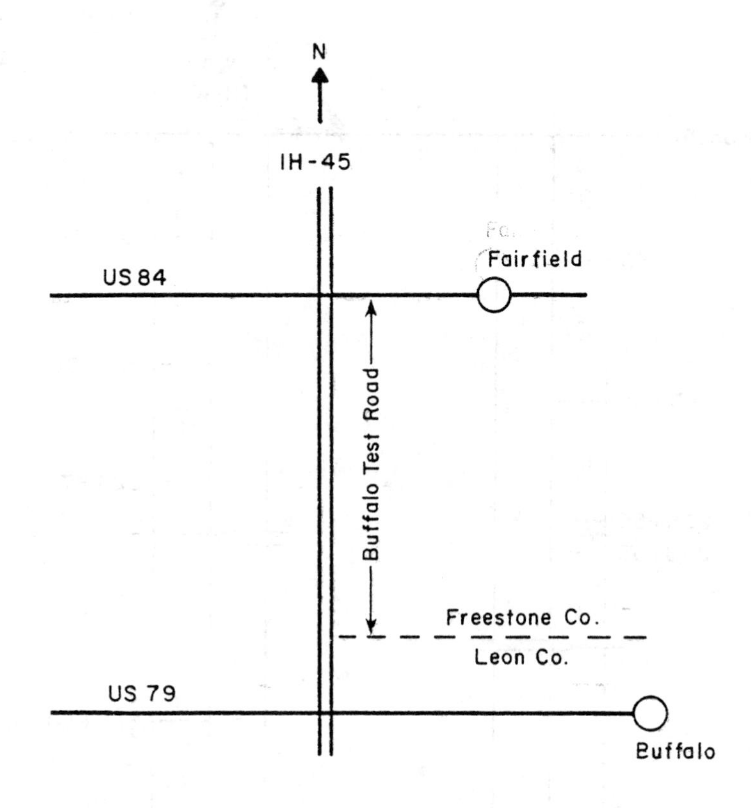

Figure 8 - Buffalo Test Road locations

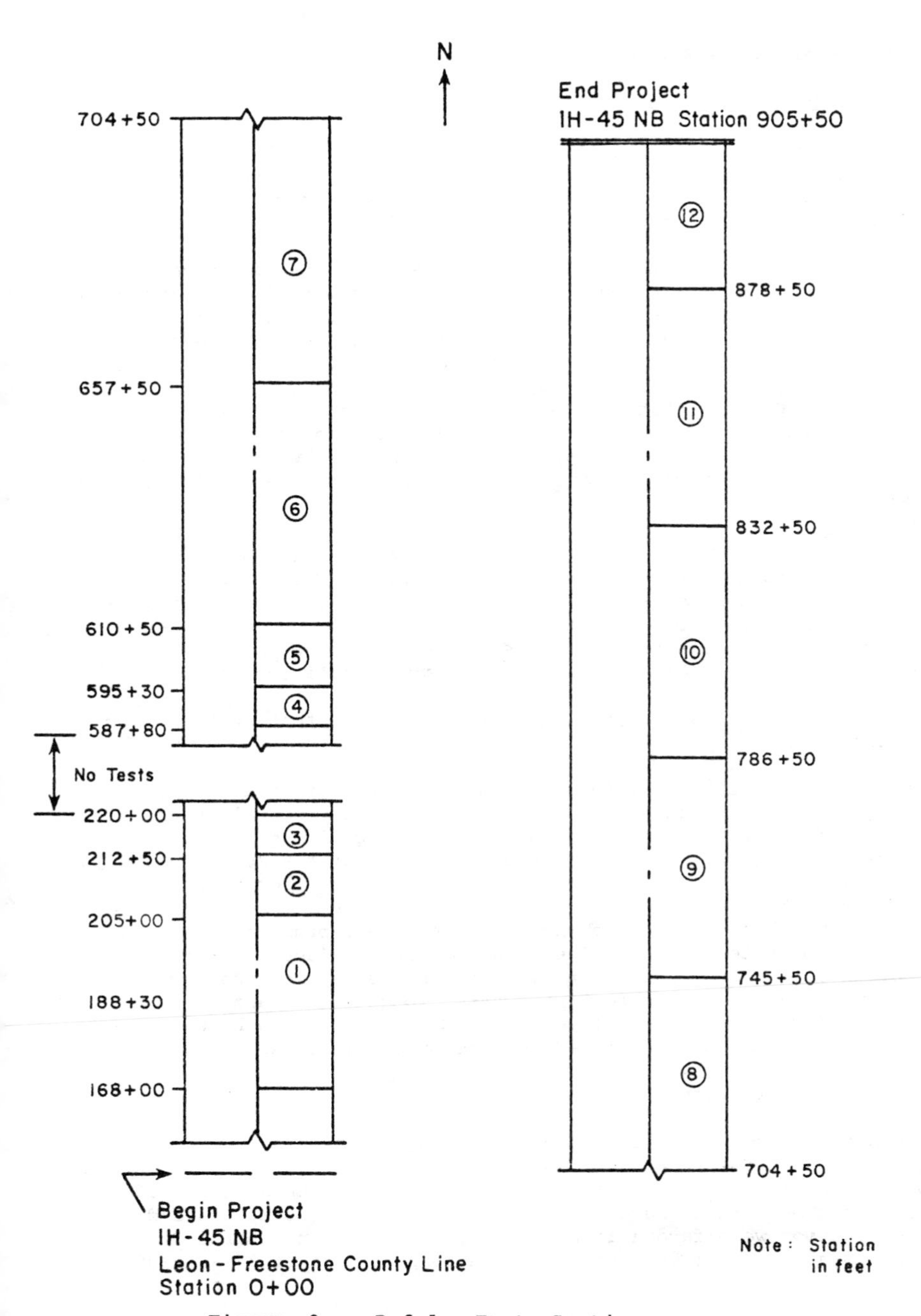

Figure 9 - Bufalo Test Sections

TEST ROAD CONSTRUCTION

El Paso-Preconstruction

Prior to construction three segments of pavement each 152.5 m (500 ft) in length were located within each test section. These sections were surveyed by photographing the 3.7 m (12 ft) wide and 152.5 m (500 ft) long pavement section prior to rehabilitation.

Photolog equipment consisted of a test vehicle equipped with a motorized 35 mm camera mounted in front of the vehicle in a vertical position over the pavement. The camera and vehicle speed were synchronized such that each photographic frame recorded pavement measuring 2.4 m by 3.7 m (8 by 12 ft) with a six inch overlap for adjacent segments. All photgraphs are on file at Texas Transportation Institute, College Station, Texas. Each photograph of the test sections was studied to determine the extent of distress present prior to construction. Distress types and levels of severity were recorded for each test section following the criteria described by Epps, et al. [8].

An index of pavement condition has been described [9] which quantifies all forms and levels of pavement distress. Based on maintenance costs, this index, or Pavement Rating Score (PRS), allows numerical comparison of pavement condition. A PRS value of 100 describes a pavement with no distress. Progressively lower PRS values describe pavement condition with more severe forms of distress.

The results of this analysis for the ten El Paso test sections appear in Table 5.

Table 5 contains the PRS values obtained by measuring all combined forms of distress present in each test section. PRS values are also shown which were obtained by measuring individual types of cracking. These cracking PRS ratings are presented such that a more precise comparison may be made between test sections for crack related distress. The asphalt-rubber interlayer is intended to reduce the rate at which cracks in the underlying pavement propagate the new asphalt concrete overlay. The "cracking PRS" values, therefore, will provide a basis for which future condition surveys can be compared. By comparing PRS values for transverse, longitudiual and alligator cracks, a measure of interlayer performance within and between test sections can be obtained based on percent original PRS.

El Paso-Construction

Asphalt-rubber interlayers were placed on June 23, 24 and 27, 1983 by International Surfacing, Inc., Phoenix, Arizona. Sections 5 to 9 were placed June 23, 1983, followed by sections 1 to 3 on June 24, 1983. Section 4 was placed June 27, 1983. Environmental conditions during construction were favorable with early morning temperatures of approximately 21C (70F) and afternoon temperatures of 37.8C (100F).

Table 5 - El Paso preconstruction pavement rating scores

Test Section	Photolog	Cracking PRS Trans.	Cracking PRS Long.	Cracking PRS Allig.	Overall PRS
1	1	7	63	65	-1
	2	70	88	70	23
	3	63	93	85	36
2	4	60	65	85	8
	5	75	93	85	48
	6	95	93	95	78
3	7	78	88	80	29
	8	73	88	80	26
	9	75	93	95	56
4	10	63	70	60	-39
	11	83	98	100	81
	12	63	70	70	-17
5	13	75	80	60	3
	14	90	78	65	28
	15	87	88	80	43
6	16	83	88	65	12
	17	83	93	80	48
	18	83	88	80	38
7	19	78	78	85	16
	20	90	88	95	63
	21	90	93	90	61
8	22	90	93	95	68
	23	78	93	95	56
	24	90	93	90	63
9	25	75	88	85	43
	26	90	70	80	28
	27	68	88	80	19
10	28	95	98	98	86

Observations and tests made during construction included the following:

I. Asphalt-rubber mixing
   A. Assuring desired rubber types were used in asphalt-rubber to be placed over selected test section locations.
   B. Proportion of asphalt and rubber.
   C. Blending time.
   D. Blending temperature.
   E. Viscosity prior to application.
   F. Sampling of asphalt and rubber.

II. Asphalt-rubber application.
   A. Asphalt-rubber spray rate.
   B. Aggregate spread rate.
   C. Asphalt-rubber cooling rate.
   D. Sampling of asphalt-rubber.

Asphalt arrived at the mixing site by highway transport where it was pumped into a storage container. Granulated rubber was shipped from the three manufacturers in 22.7 kg or 27.2 kg (50 or 60 lb) bags.

Results of observations and tests performed during mixing of the asphalt-rubber appear in Table 6. Note that the field viscosity of the asphalt-rubber blend appears to depend on rubber content as shown in Table 6 and plotted in Figure 10. Note that the type of rubber affects the viscosity of the blend as shown in Figure 10. Viscosity tests were performed using a portable Haake rotational viscometer on samples of asphalt-rubber obtained directly from the distributor truck approximately 50 minutes after all rubber had been added to the truck.

Rubber Type C, in addition to generating the lowest asphalt-rubber viscosity, relationship, also caused a considerable volume increase in the blend as mixing progressed. This was manifested in overflows of asphalt-rubber from the top hatch of the 17,055 l (4500 gallon) distributor truck for test section mixes 6 and 7. The overflows occurred during routine pumping of the blend after approximately 8717 l (2300 gallons) had been loaded. Overflow was avoided for the third blend containing Rubber C by loading the first half of the blend at a slower rate. Moisture contained in the rubber is thought to be the cause of this adverse reaction and may be related to the cryogenic processing technique.

Buffalo-Preconstruction

Eight sections of pavement each approximately 0.80 lane-mile in length were selected to receive the various asphalt-rubber blends shown in Figure 6. Four additional pavement sections, each 229 m (750 ft) in length, were selected as control sections. Three segments of pavement each 153 m (500 ft) in length were selected in each of the eight test sections for photolog surveys. The entire length of the control sections were photologged.

Table 6 - El Paso mixing observations an test results

Test Section	Beginning Date	Time of Day	Time Req'd to Fill Truck w/ Blend, min.	Time Between Full Truck & Application, min.	Temp, F Prior to Application	Viscosity Prior to Application, poises	Rubber Type	Rubber Content, Percent
1	6/24/83	4:35am	40	105	320	20	A	24
2	6/24/83	5:20am	40	95	390	9	C	22
3	6/24/83	6:02am	53	90	320	35	B	26
4	6/27/83	11:40am	35	110	338	18	B	22
5	6/23/83	5:25am	55	85	340	15	A	22
6	6/23/83	6:25am	55	90	330	15	C	26
7	6/23/83	11:20am	30	160	345	10	C	24
8	6/23/83	1:15pm	30	135	325	23	A	26
9	6/23/83	1:50pm	30	125	330	25	B	24

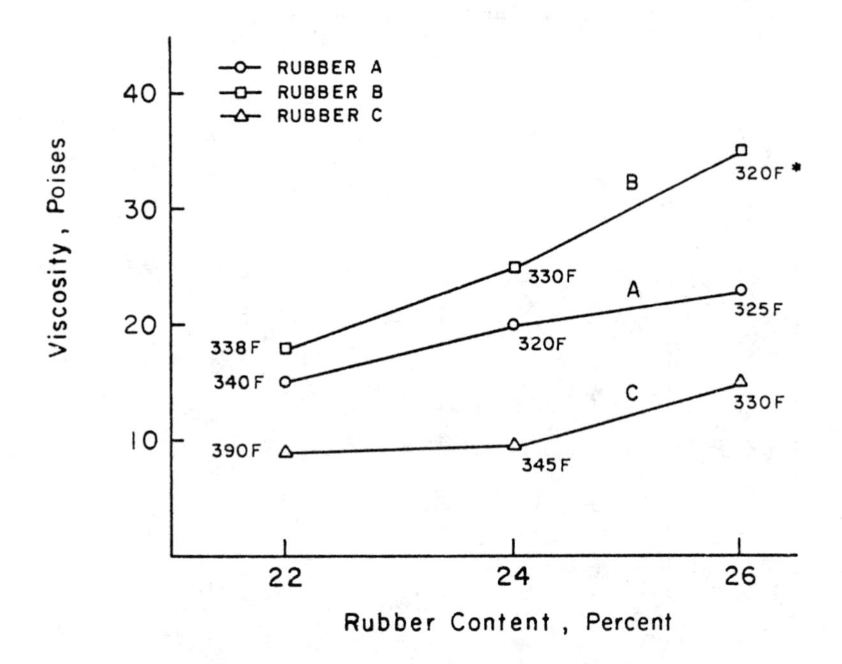

Figure 10 - El Paso asphalt-rubber viscosity

Buffalo-Construction

Asphalt-rubber was placed over test sections August 20, 21 and 22, 1984 by Arizona Refining Company, Phoenix, Arizona. Environmental conditions during construction were favorable with early morning temperatures of approximately 21C (70F) and afternoon temperatures approaching 38C (100F).

Blending of asphalt and Sundex 790 at 6 percent Sundex by blend volume was accomplished prior to blending with rubber. Pre-blending of asphalt-rubber was accomplished as on the El Paso project prior to pumping the blend into distributor trucks. Here the asphalt-rubber blend remained in the trucks for the desired digestion period prior to application.

Digestion was varied as a control variable in this experiment as explained previously for laboratory prepared mixes. Two levels of digestion were achieved. "Low" digestion describes blends of 2 to 2 3/4 hours. "High" digestion describes blends of 16 to 16 1/2 hours.

Rubber concentrations of 18 and 22 percent by weight of the blend were used.

Materials application rates were monitored during construction for each section. Results of this testing is shown in Table 7.

TABLE 7 -- Materials application rates for Buffalo Test Road.

Section	Binder, lsm (gsy)	Aggregate, sm/cm (sy/cy)
1	2.63 (0.58)	104 (95)
2	2.58 (0.57)	98 (90)
3	None	None
4	None	None
5	2.49 (0.55)	88 (80)
6	2.58 (0.57)	84 (77)
7	2.54 (0.56)	86 (79)
8	2.58 (0.57)	84 (77)
9	2.36 (0.52)	82 (75)
10	2.67 (0.59)	88 (80)
11	2.45 (0.54)	85 (78)
12	2.54 (0.56)	86 (79)

Locations of photologs within test sections are permanently marked using raised reflective pavement buttons positioned on the right shoulder of the northbound lane. Precise location of photologs for future condition surveys is therefore possible by reference to these pavement markers.

LABORATORY TESTS

Three laboratory tests were used to evaluate physical properties of asphalt-rubber blends prepared in the field and in the laboratory.

The mixer used to blend asphalt and rubber in the laboratory also served as a rotational viscometer to evaluate rheological characteristics during blending.

Each device used has been has been reported in other research where application of each has been demonstrated [4, 5, 6, 7, 11, 12].

Torque Fork Mixer

A laboratory mixer of this type was first used for asphalt-rubber blending in 1977 [4]. The system consists of a constant speed motor with stirrer assembly which is capable of recording torque changes as load varies on the stirrer. The resulting apparatus is a rotational viscometer which can measure relative changes in fluid viscosity during mixing.

Haake Viscometer

A Haake portable rotational viscometer model VT-02 was used in the field and laboratory to determine the viscosity of both laboratory and field mixed asphalt-rubber blends. The Haake is a simple device which measures viscosity by the same principle as the torque fork mixer, except changes in torque are monitored by deflection of a calibrated spring rather than by increases in electrical current.

Laboratory blended asphalt-rubber was tested by immersing the viscometer cup in the blend after digestion in the Torque Fork Mixer. The procedure for obtaining viscosity data was as for field prepared blends.

Force Ductility

The force ductility test is a modification of the asphalt ductility test [10]. The test has been described [6, 11] as a means to measure tensile load-deformation characteristics of asphalt and asphalt-rubber binders.

The test is performed as described by ASTM D113 [10] with certain changes. The principal alteration of the apparatus consists of adding two force cells in the loading chain.

Double Ball Softening Point

This test is based on a concept proposed by Krchma [12] for characterization of asphalts. It is a modified version of the ASTM Ring and Ball Softening Point Test [13].

The double ball softening point test apparatus consists of two 0.95 cm (3/8 in) diameter stainless ball bearings cemented together with the test material. One of the ball bearings is fixed to the ring holder of the standard ring and ball assembly, the other ball is suspended from the first by the test material.

LABORATORY TEST RESULTS

Force ductility and double ball softening point were performed with asphalt-rubber mixes prepared in the field and in the laboratory.

Laboratory mixes were prepared using the Torque Fork Mixer. All testing was performed using a completely random sequence.

Torque Fork Mixer

All laboratory blends of asphalt-rubber were prepared in the torque fork as previously described. Twenty-seven blends were prepared for all combinations of variables as shown in Figure 5. Output from the torque fork is in terms of millivolts of electromotive force required to maintain a stirring speed of 500 rpm.

Results of testing indicate asphalt-rubber viscosity increases with time and temperature as shown in Figure 11. The initial rapid increase in viscosity is

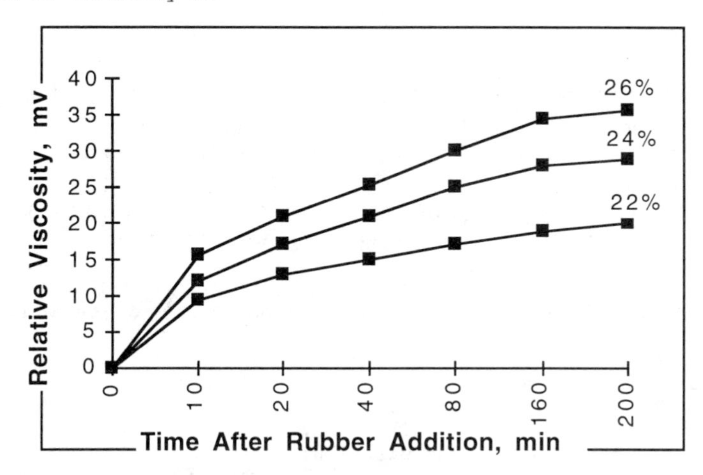

FIG. 11 -- Torque fork viscosity during mixing.

represented by the low level of digestion, with an approximate constant increase in viscosity after high digestion time.

Solvent Extraction of Rubber

A test was devised to extract asphalt from asphalt-rubber mixes using a procedure described in ASTM D2172 Method B to determine if rubber disintegrates while digested at various levels with asphalt. The results are shown for Buffalo and El Paso asphalt-rubber blended in the laboratory with 22 percent Type A rubber under the conditions shown in Figure 12:

These results support data previously presented which indicates that as digestion level increases, solid rubber disintegrates, leaving less rubber by weight in the mix.

Size Exclusion Chromatography

A gel permeation chromatography (GPC) test [15] was used to identify changes between original asphalt and asphalt after mixing with rubber to determine if digestion of rubber in asphalt causes a molecular

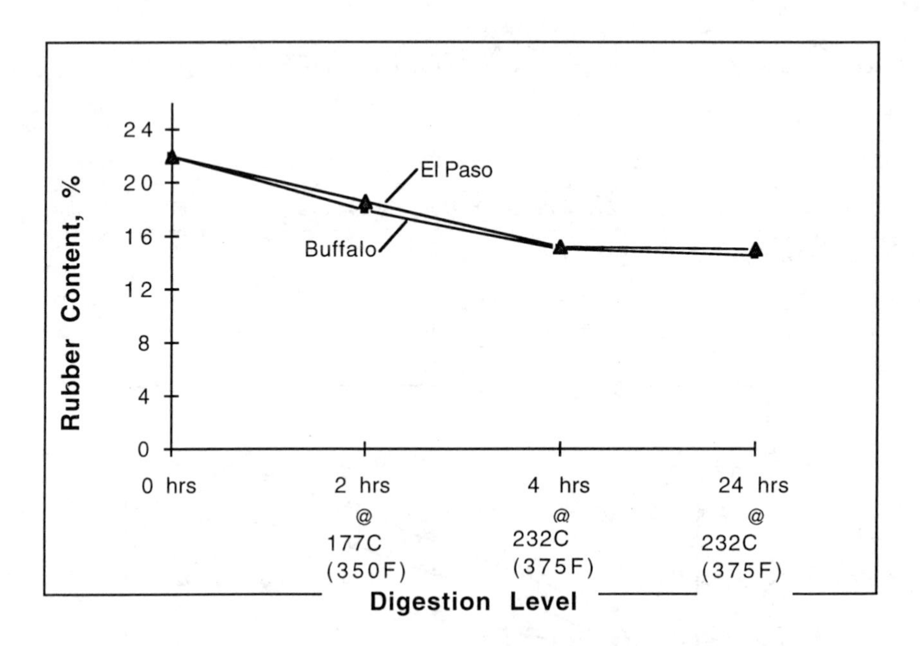

FIG. 12 -- Loss of solid rubber with digestion.

change in the asphalt-rubber blend. The GPC test provided data regarding the molecular weight distribution of asphalt before and after digestion with rubber. The results shown in Figure 13 depict a shift in the molecular weight distribution after digestion. This shift in molecular weight of the asphalt after digestion indicates that alterations have occurred in the asphalt during the digestion process as noted by an increase in high and low molecular weight materials. This means that as digestion continues, some rubber may be lost to the asphalt fraction of the asphalt-rubber mixture. This has been shown by extraction results above and by the increase in both high and low molecular weights as shown by GPC results in Figure 13.

Force Ductility

Seven test responses appearing in Table 8 and diagrammed in Figure 14 were measured from results of each force ductility test for both field prepared and laboratory prepared mixes. Multiple analysis of variance (ANOVA) techniques were employed to determine whether differences in material properties could be measured between the various factors investigated.

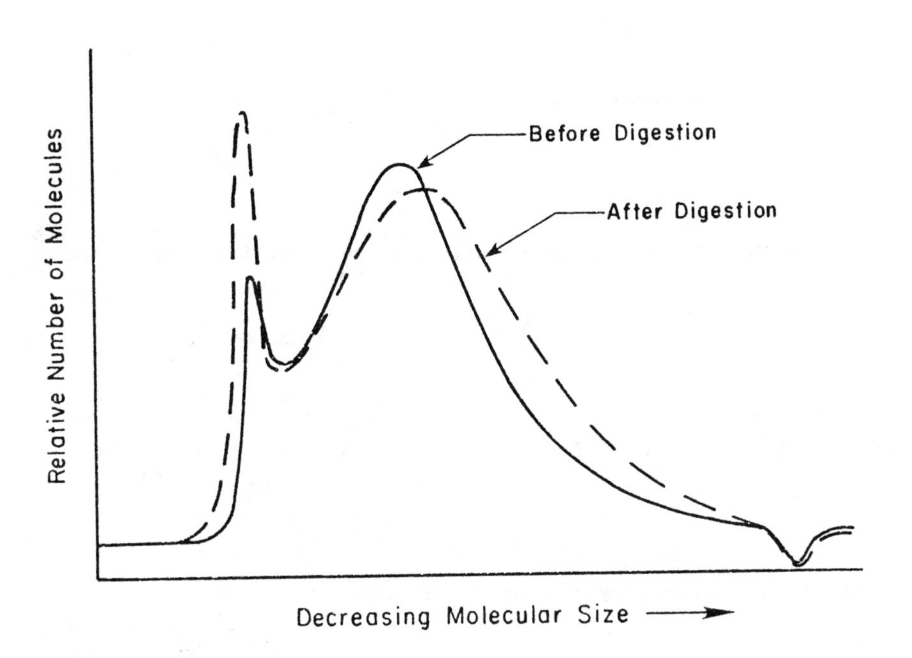

FIG. 13 -- Gel permeation chromatography results

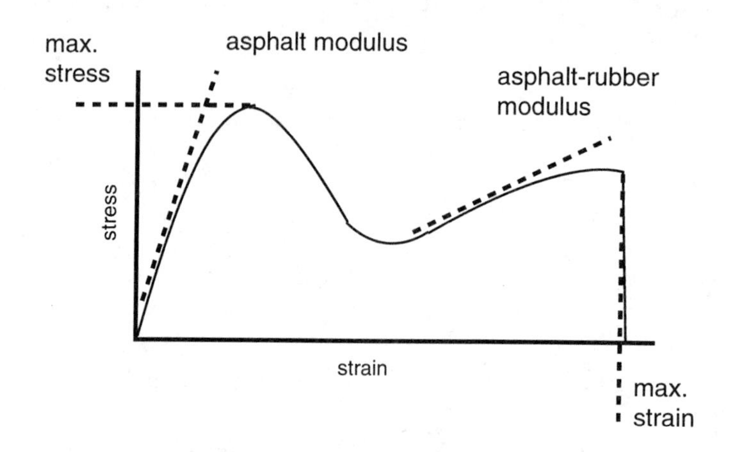

FIG. 14 -- Characteristic force-ductility curve for asphalt-rubber

El Paso Mixes

A summary of ANOVA results from lab mixes appear in Table 8. Check marks in the tables indicate no significant difference at α = 0.05. Table 9 provides similar information for results of field prepared asphalt-rubber mixes.

TABLE 8 -- Laboratory prepared asphalt-rubber for El Paso Test Road.

Factor	Max. Eng Stress	Max. Eng Strain	Max. True Stress	Max. True Strain	Curve Area	Asphalt Modulus	Asph-Rub Mod
Rubber	•	•	•	•	•	•	•
Concentration	•					•	
Digestion		•		•		•	•

TABLE 9 -- Field prepared asphalt-rubber for El Paso Test Road.

Factor	Max. Eng Stress	Max. Eng Strain	Max. True Stress	Max. True Strain	Curve Area	Asphalt Modulus	Asph-Rub Mod
Rubber Concentration		•		•			•

After judging differences between control variables significant at α = 0.05, the Newman-Keuls [14] multiple comparison procedure was used to judge which treatment means contributed to the significant ANOVA results. Newman-Keuls analysis was applied when ANOVA indicated significance as shown in Tables 8 and 9. The results of the Newman-Keuls analysis appear in Tables 10 and 11 for laboratory and field prepared mixes.

TABLE 10 -- Significant factors for laboratory prepared asphalt-rubber for El Paso Test Road.

Factor	Max. Eng Stress	Max. Eng Strain	Max. True Stress	Max. True Strain	Curve Area	Asphalt Modulus	Asph-Rub Mod
Rubber							
A	su	s	s	s	s	su	s
B	s	u	s	u	s	s	su
C	u	t	u	t	u	u	u
Concentration							
22						u	
24						s	
26						s	
Digestion							
L		s		s		u	s
M		s		s		s	su
H		u		u		s	u

* Factors with same symbol are not significantly different at $\alpha = 0.05$

TABLE 11 -- Significant factors for field prepared asphalt-rubber for El Paso Test Road.

Factor	Max. Eng Stress	Max. Eng Strain	Max. True Stress	Max. True Strain	Curve Area	Asphalt Modulus	Asph-Rub Mod
Rubber							
A							s
B							u
C							su
Concentration							
22							
24		su		u			
26		s		s			
		s		s			

* Factors with same symbol are not significantly different at $\alpha = 0.05$

Buffalo Mixes

ANOVA results have been summarized in Tables 12 and 13. A bullet indicaes no significant difference in factors at $\alpha = 0.05$.

TABLE 12 -- Laboratory prepared asphalt-rubber for Buffalo Test Road.

Factor	Max. Eng Stress	Max. Eng. Strain	Max. True Stress	Max. True Strain	Curve Area	Asphalt Modulus	Asph-Rub Mod
Concentration						•	•
Digestion	•	•	•	•	•	•	•
Dig x Conc	•		•		•	•	

TABLE 13 -- Field Prepared asphalt-rubber for Buffalo Test Road.

Factor	Max. Eng Stress	Max. Eng. Strain	Max. True Stress	Max. True Strain	Curve Area	Asphalt Modulus	Asph-Rub Mod
Concentration	•	•		•			•
Digestion	•	•	•	•	•	•	•
Replicate							

Note that for Buffalo field mixes, a significant difference between replicates is rejected at alpha ≤ 0.05. However, significance between digestion periods and between rubber concentrations is detected using parameters MES, MTS, and ARM.

Different digestion periods for laboratory mixed asphalt - rubber produces highly significant ANOVA results for each parameter tested as shown in Table 12.

Asphalt-rubber modulus (ARM) indicates significance for concentration and digestion. This is the only parameter which identifies significance for both of these factors in both field prepared and laboratory prepared mixes.

Newman-Keuls analysis was used for Buffalo mixes to determine which levels of factors were significantly different as described by ANOVA. Results are shown in Table 14 such that trends present in the data can be more easily observed. The results of the Newman-Keuls analysis for laboratory prepared mixes appear in Table 15.

TABLE 14 -- Significant factors for laboratory prepared asphalt-rubber for Buffalo Test Road.

Factor	Max. Eng Stress	Max. Eng. Strain	Max. True Stress	Max. True Strain	Curve Area	Asphalt Modulus	Asph-Rub Mod
Concentration							
18	s	u		u			s
22	u	s		s			u
Digestion							
L							
H	u	s	u	s	u	u	u
	s	u	s	u	s	s	s

* Factors with same symbol are not significantly different at $\alpha = 0.05$

TABLE 15 -- Significant factors for field prepared asphalt-rubber for Buffalo Test Road.

Factor	Max. Eng Stress	Max. Eng. Strain	Max. True Stress	Max. True Strain	Curve Area	Asphalt Modulus	Asph-Rub Mod
Concentration							
18						u	s
22						s	u
Digestion							
L							
M	t	s	u	s	u	t	s
H	u	u	u	u	u	u	u
	s	u	s	t	s	s	s

* Factors with same symbol are not significantly different at $\alpha = 0.05$

Double Ball Softening Point

ANOVA results for laboratory mixes indicate no significant differences for main factors or interactions at an α level of 0.05. The test is sensitive to rubber type at an α level of 0.18 and to the rubber-digestion interaction at $\alpha = 0.14$.

ANOVA results for field mixed material indicate sensitivity for differences in rubber concentration at $\alpha = 0.18$ and to rubber-concentration interaction at $\alpha = 0.06$.

It is possible that imprecision in this test is responsible for the lack of discrimination measured rather than because the materials have the same properties. Much difficulty was experienced when molding asphalt-rubber when conducting this test.

SUMMARY

The force ductility test appears to be capable of discriminating between different types of asphalt-rubber blends. Those independent variables able to measure differences in factors for laboratory prepared blends for each test road are shown in the tables below:

TABLE 16 -- Discriminating responses for laboratory prepared asphalt-rubber on the El Paso Test Road.

Test Response	Rubber Concentration	Digestion Level	Rubber Type
Failure Stress			•
Failure Strain		•	•
Asphalt Modulus	•	•	•
Asph-Rubb Mod		•	•
Curve Area			•
			•

TABLE 17 -- Discriminating responses for laboratory prepared asphalt-rubber on the Buffalo Test Road.

Test Response	Rubber Concentration	Digestion Level
Failure Stress		•
Failure Strain		•
Asphalt Modulus	•	•
Asph-Rubb Mod	•	•
Curve Area		•

TABLE 18 -- Discriminating responses for field prepared asphalt-rubber on the El Paso Test Road .

Test Response	Rubber Concentration	Digestion Level	Rubber Type
Failure Stress			
Failure Strain	•		
Asphalt Modulus			
Asph-Rubb Mod			
Curve Area			•

TABLE 19 -- Discriminating responses for field prepared asphalt-rubber on the Buffalo Test Road.

Response	Rubber Concentration	Digestion Level
Failure Stress		•
Failure Strain	•	•
Asphalt Modulus		•
Asph-Rubb Mod		•
Curve Area	•	•

CONCLUSIONS

1. Asphalt-rubber binders can be formulated for spray applications using various combinations of asphalt, ground tire rubber, and extender oils. Both ambiently ground and cryogenically ground rubber can be used. Rubber concentrations in this study were varied from 18 to 26 percent by weight of the blend.

2. Extraction of asphalt from asphalt-rubber mixtures indicates solid rubber disintegrates in the asphalt-rubber as digestion proceeds. This disintegration appears to be time and temperature driven.

3. Size exclusion chromatography results indicate asphalt-rubber mixtures contain more high molecular size and low molecular size molecules than the original parent asphalt. This addition of both

high and low molecular sizes is evidence of modification of the original asphalt by the rubber.

4. The force-ductility test yields seven parameters which are sensitive to changes in asphalt-rubber properties caused by varying rubber type, rubber concentration, and digestion level.

5. The force-ductility test produces a true stress-strain curve for asphalt-rubber mixtures which has two characteristic linear portions. The slope of the linear portion of the curve during initial loading approximates the modulus of the asphalt cement. The slope of the linear portion of the curve during later stages of loading may measure a composite "asphalt-rubber modulus". The slopes of both portions of the stress-strain curve appear to be a function of rubber content and digestion level and may be useful for characterizing asphalt-rubber binders.

6. Digestion and rubber content generally counteract each other with respect to asphalt-rubber properties. Digestion generally increases failure strain and decreases failure stress and modulus. Rubber content generally decreases failure strain but increases failure stress and modulus. Therefore, an increase or decrease in tensile properties may be achieved with asphalt-rubber mixtures by varying digestion conditions and/or rubber content. However, as digestion proceeds, evidence suggests an apparent loss of rubber content. Therefore, physical properties of asphalt-rubber blends are digestion time dependent.

7. Physical properties of some field prepared asphalt-rubber mixtures appear similar to mixtures prepared in the torque fork laboratory mixer/viscometer. However, low level field digestion corresponds to a higher level of digestion in the laboratory.

8. The double-ball softening point test was sensitive to changes in certain mixtures, but insensitive to others. The non-homogeneous nature of certain asphalt-rubber blends is believed to be the cause for this inconsistency. Because of this inconsistency, this test may be of questionable utility for characterization purposes.

REFERENCES

[1] Shuler, T. S., Pavlovich, R. D. and Epps, J. A. "Field Performance of Rubber Modified Asphalt Paving Materials," Transportation Research Board, Washington, D.C., January, 1985.

[2] Way, G. B., "Prevention of Reflective Cracking at Minnetonka - East (1979 Addendum Report)," Report ADOT-RS-15(130), Arizona DOT, August, 1979.

[3] Oliver, J. W. H., "A Critical Review of the Use of Rubbers and Polymers in Bitumen Bound Paving Materials," Interim Report AIR-1037-1, Australian Road Research Board, Victoria, 1977.

[4] Pavlovich, R. D., Shuler, T. S., and Rosner, J. C., "Chemical and Physical Properties of Asphalt-Rubber," Report ADOT-RS-15(133), Arizona DOT, November, 1979.

[5] Green, E., and Tolonen, W. J., "Chemical and Physical Properties of Asphalt-Rubber Mixtures," Report ADOT-RS-14(162), ADOT, July, 1977.

[6] Shuler, T. S. and Hamberg, D. J., "A Rational Investigation of Asphalt-Rubber Properties," University of New Mexico Engineering Research Institute, Albuquerque, August, 1980.

[7] Jimenez, R. A., "Testing Methods for Asphalt-Rubber," Report ADOT-RS-15(164), ADOT, January, 1978.

[8] Epps, J. A., Meyer, A. H. Larrimore, I. E., Jr., and Jones, H. L. "Roadway Maintenance Evaluation User's Manual," Texas Transportation Institute Research Report 151-2, September, 1974.

[9] Epps, J. A. Larrimore, I. E., Jr., Meyer, A. H., Cox, S. G., Jr., Evans, J. R., Jones, H. L., Mahoney, J., Wootan, C. V., and Lytton, R. L., "The Development of Maintenance Management Tools for Use by the Texas State Department of Highways and Public Transportation," Texas Transportation Institute Research Report 151-4F, September, 1976.

[10] Annual Book of ASTM Standards, 1984, Section 4, Volume 04.03, "Standard Test Method for Ductility of Bituminous Materials," Designation D113-79.

[11] Anderson, D. I., Wiley, M. L., "Force-Ductility An Asphalt Performance Indicator," AAPT, Vol. 45, 1976.

[12] Krchma, L. C., "Asphalt Consistency Control and Characterization by Flow Temperature," AAPT Volume 36, 1967.

[13] Annual Book of ASTM Standards, 1984, Section 4 , Volume 04.04, "Standard Test Method for Softening Point of Bitumen (Ring and Ball Apparatus)," Designation D36-76.

[14] Anderson, V. L., and McLean, R. A., Design of Experiments A Realistic Approach, Marcel Dekker, Inc., New York, 1974.

[15] Annual Book of ASTM Standards, 1984, Section 8, Volume 08.03, "Standard Test Method for Molecular Weight Distribution of Certain Polymers by Liquid Size-Exclusion Chromatography (Gel Permeation Chromatography, GPC)," Designation D3593.

[16] Soil Interpretation Records of Edward I. Janak, Jr., Soil Scientist, Soil Conservation Service, from personal correspondence, Corsicana, Texas, December, 1983.

[17] Annual Book of ASTM Standards, 1983, Section 9, Volume 09.01, "Standard Methods for Rubber Products-Chemical Analysis," Designation D297-81.

Kent R. Hansen[1], and Gary Anderton[2]

A LABORATORY EVALUATION OF RECYCLED TIRE RUBBER IN HOT-MIX ASPHALT PAVING SYSTEMS

REFERENCE: Hansen, K. R. and Anderton, G., **"A Laboratory Evaluation of Recycled Tire Rubber in Hot-Mix Asphalt Paving Systems,"** Use of Waste Materials in Hot-Mix Asphalt, ASTM STP 1193, H. Fred Waller, Ed., American Society for Testing and Materials, Philadelphia, 1993.

Abstract: This report digests individual studies by separate researchers which compared and evaluated: (1) the physical properties and aging characteristics of the asphalt-rubber and asphalt cement binders; (2) Hveem and Marshall mix design methods, and permanent deformation, low temperature cracking and fatigue characteristics for dense graded asphalt-rubber concrete (ARC) and asphalt concrete mixes; and (3) open-graded asphalt-rubber concrete friction courses.

Many of the physical properties investigated in this research program were significantly enhanced by the addition of ground recycled rubber to the asphalt cement binder. These asphalt-rubber mixtures were found to act quite differently from traditional, unmodified asphalt mixtures. However, these results indicate that improved pavement performance can be achieved with asphalt-rubber binder if the design and construction criteria are changed to reflect the unique physical characteristics imparted by these special binders.

Keywords: asphalt-rubber, permanent deformation, low temperature cracking, fatigue cracking, aging, mix design, tensile creep, dense-graded, open-graded

---

In November of 1989, the U.S. Army Corps of Engineers Waterways Experiment Station and the Asphalt Rubber Producers Group signed a Cooperative Research and Development Agreement which marked the beginning of a two-year joint research study on asphalt-rubber. This report digests the results obtained from the two-year asphalt-rubber research study. Individual studies of differing research areas were conducted by several agencies including: the U.S. Army Engineer Waterways Experiment Station's (WES) Pavement Systems Division, the University of Nevada-Reno (UNR), the University of Arizona (UA),

---

[1] Senior Design Engineer, International Surfacing, Inc., 6751 W. Galveston, Chandler, AZ 85226

[2] Civil Engineer, U.S. Army Engineer Waterways Experiment Station, Vicksburg, MS 39108-6119

International Surfacing, Inc. (ISI) and Crafco, Inc. Detailed reports of these individual studies are documented in the Technical Reports listed below:

Volume 1	-	Summary of Research on Asphalt-Rubber Binders and Mixes
Volume 2	-	Physical Properties and Aging Characteristics of Asphalt-Rubber Binders
Volume 3	-	Tensile Creep Comparison of Asphalt Cement and Asphalt-Rubber Binders
Volume 4	-	Comparison of Mix Design Methods for Asphalt-Rubber Concrete Mixtures
Volume 5	-	Permanent Deformation Characteristics of Recycled Tire Rubber Modified and Unmodified Asphalt Concrete Mixtures
Volume 6	-	Low Temperature Cracking Characteristics of Ground Rubber and Unmodified Asphalt Concrete Mixtures
Volume 7	-	Fatigue of Asphalt and Asphalt-Rubber Concretes
Volume 8	-	Asphalt-Rubber Open-Graded Friction Courses

Copies of these Technical Reports may be obtained by contacting:

Asphalt Rubber Producers Group
312 Massachusetts Ave., NE
Washington, D.C. 20002

Materials

Binders

Binders described in this report are abbreviated as follows:

AC-5	Witco AR 1000 (AC-5) Asphalt Cement.
AC-20	Witco AC-20 Asphalt Cement.
AC-40	Witco AC-40 Asphalt Cement.
AC-5RE	79% Witco AR 1000(AC-5) Asphalt Cement, 5% San Joaquin 1200S Extender Oil and 16% Baker IGR 24 Rubber.
AC-5R	83% Witco AR 1000 (AC-5) Asphalt Cement and 17% Baker IGR-24 Rubber.
AC-20R	84% Witco AC-20 Asphalt Cement and 16% Baker IGR-24 rubber.

Aggregate

The aggregate was obtained from Granite Rock Co., Watsonville, California. The gradation was chosen to meet ASTM Standard Specification for Hot Mixed, Hot Laid Bituminous Pavements (ASTM D 3515) 1/2-inch dense mixture, Nevada Type 2 and California 1/2-inch medium specifications.

BINDER TESTING

Physical Properties and Aging Characteristics of Asphalt-Rubber Binders

Asphalt cement and asphalt-rubber binders were evaluated in report Volume 2, "Physical Properties and Aging Characteristics of Asphalt-Rubber Binders"[1], using the following tests:

- Absolute Viscosity at 60°C - ASTM Test Method for Viscosity of Asphalts by Vacuum Capillary Viscometer (ASTM D 2171)

- Kinematic Viscosity at 135°C - asphalt cement only - ASTM Test Method for Kinematic Viscosity of Asphalts (Bitumens) (ASTM D 2170)
- Brookfield Viscosity at 90°C to 135°C - ASTM Method of Testing for Rubberized Tar (ASTM D 2994)
- Needle Penetration and Cone Penetration at 4°C and 25°C - ASTM Standard Test Method for Penetration of Bituminous Materials (ASTM D 5) and ASTM Standard Test Method for Cone Penetration of Lubricating Grease (ASTM D 217), respectively.
- Ductility at 4°C & 25°C - ASTM Test Method for Ductility of Bituminous Materials (ASTM D 113)
- Softening Point - ASTM Test Method for Softening Point of Bitumen (Ring and Ball Apparatus) (ASTM D 36)
- Resilience ASTM Methods of Testing Joint Sealants, Hot-Poured, for Concrete and Asphalt Pavement (ASTM D 3407)

Binders were also evaluated after aging using ASTM Standard Test Method for Effect of Heat and Air on Asphaltic Materials (TFOT) (ASTM D 1754), Thin Film Oven Test (TFOT) and Weatherometer (Federal Specification SS-S-00200E). Binders were tested using the following test methods before and after aging:

- Absolute Viscosity at 60°C (ASTM D 2171)
- Needle Penetration and Cone Penetration at 25°C (ASTM D 5 and D 217, respectively)
- Softening Point (ASTM D 36)
- Weight Loss

Figure 1 shows typical temperature viscosity relationships for asphalt-rubber and unmodified asphalt cements. These and other viscosity test results demonstrate the conclusion that the addition of

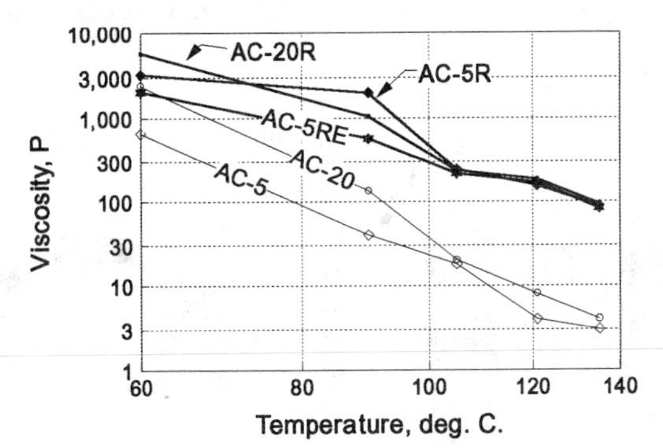

Figure 1 -- Absolute and Brookfield Viscosities

16 to 17 percent ground recycled tire rubber to an asphalt cement will increase the binder viscosity by 100 to 2000 percent, depending upon the test method and test temperature.

The viscosity tests also show that differing grades of asphalt-rubber binders produced with similar dosage levels of the same rubber have very similar viscosities between 100°C and 135°C. This indicates that above about 100°C, the viscosity of the binder is controlled by the rubber and below 90°C, the base asphalt cement has a significant influence on binder viscosity.

Needle penetration test results for four of the binders are shown in Figure 2. These data illustrate that the addition of recycled tire rubber can: (1) improve low-temperature binder properties as indicated by comparing the 4°C pen for the AC-20 and AC-20R; and (2) reduce overall temperature susceptibilities as indicated by the difference between the 4 and 25°C penetration tests.

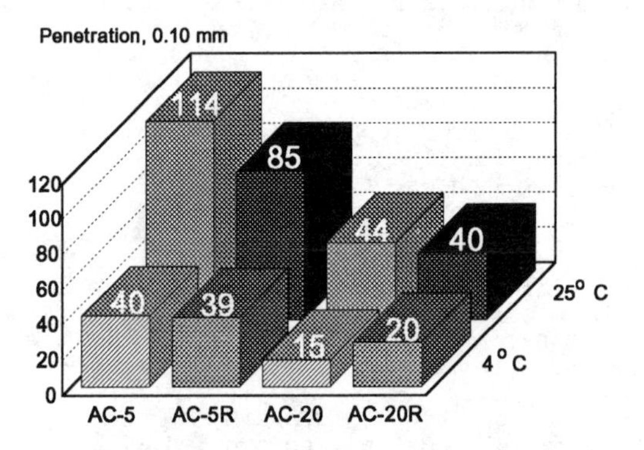

Figure 2 -- Needle Penetration

Softening point test results for four binders are shown in Figure 3. These tests show softening points are increased by approximately 11 to 17°C by the addition of 16 to 17 percent recycled tire rubber. It is important to note that the AC-5R has a higher softening point than the AC-20. This testing indicates that asphalt-rubber concrete pavements should be less susceptible to traffic-induced deformation distress at high pavement temperatures. This may also be true when comparing the AC-5RE to the AC-20.

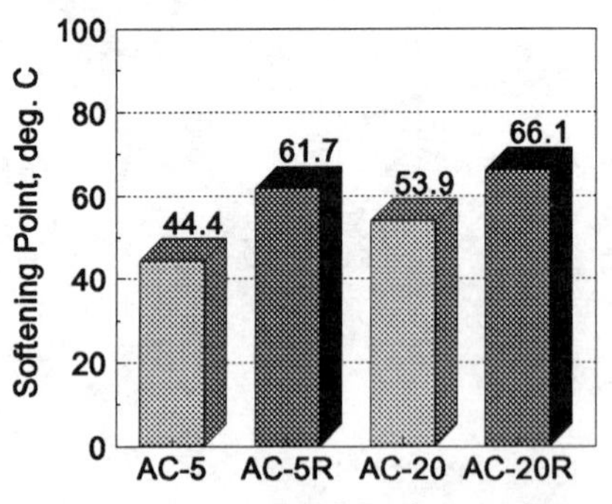

Figure 3 -- Softening Point

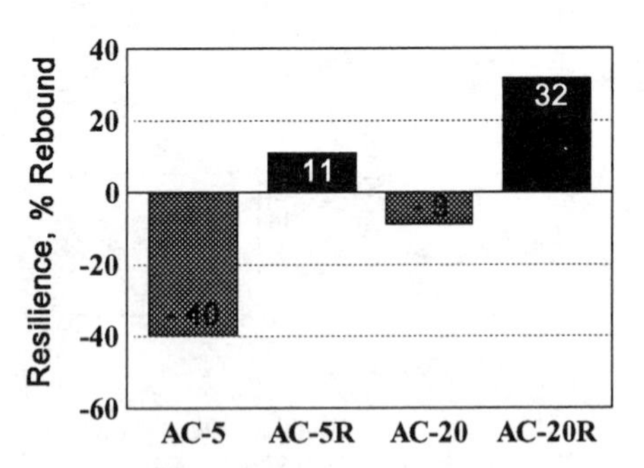

Figure 4 -- Resiliency

The resiliency data shown in Figure 4 measures the ability of a binder to recover from a set strain at 25°C. This test shows asphalt-rubber binders have significantly higher elastic recovery potentials than unmodified asphalt cement binders. This test also indicates that asphalt-rubber concrete mixes should show improved resistance to high temperature deformation when compared to unmodified mixes.

Viscosity and 25°C penetration test results on thin film oven test (TFOT) aged samples are shown in Figure 5. The viscosity tests illustrate that hot-mix asphalt plant aging of asphalt-rubber binders, with the exception of the AC-5RE, is about 50 percent less than asphalt cement binders. The penetration data show improved plant aging resistance for all asphalt-rubber binders.

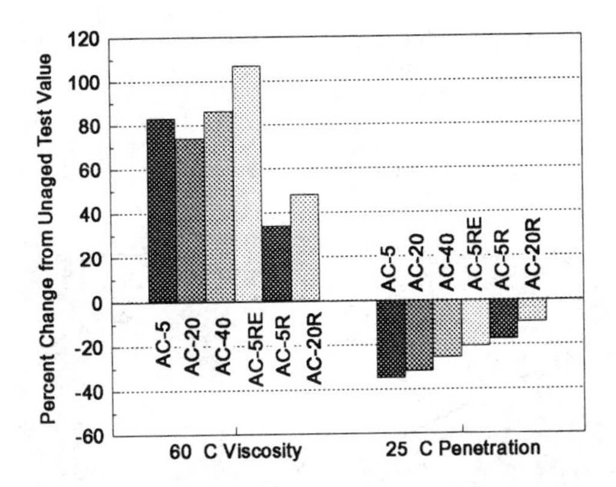

Figure 5 -- Thin Film Oven Aging

Asphalt-rubber binders had higher weight losses after thin film oven test aging when compared to the asphalt cement binders, but the amount of weight loss did not appear to significantly affect other aging properties.

Figure 6 presents the results of viscosity and penetration tests on samples aged in the weatherometer. This accelerated aging subjects the specimens to heat, ultraviolet light and moisture to simulate environmental aging. Viscosity data show all asphalt-rubbers tested, except the AC-5RE, exhibit reduced environmental aging. The penetration test shows improved aging resistance for all asphalt-rubber binders when compared to the base asphalt cement.

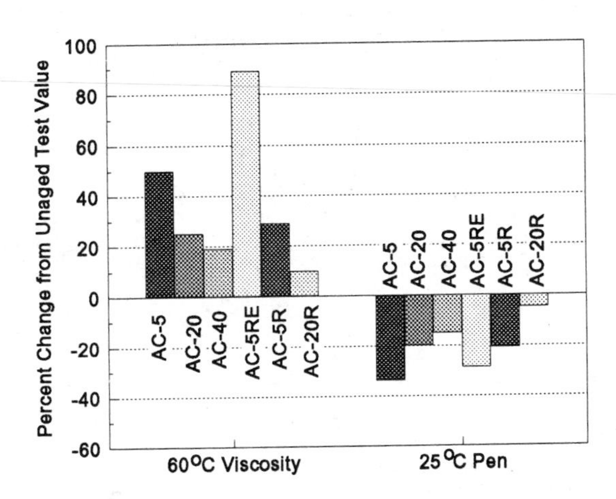

Figure 6 --Weatherometer Aging (8 Days)

Tensile Creep Comparison of Asphalt Cement and Asphalt-Rubber Binders

Binders were evaluated in Volume 2, "Tensile Creep Comparison of Asphalt Cement and Asphalt-Rubber Binders" [2], using creep test procedures reported by Coetzee and Monismith [3]. In brief, samples are tested in a modified ductility bath where a load is applied using a dead weight and pulley system. The load was selected to obtain a strain of 20 to 40 percent at 1,000 seconds. Samples were tested at -5.5, 4, 13 and 25°C. Stiffness moduli were calculated, and plotted for each test. A typical stiffness modulus verses temperature plot is presented in Figure 7 (the shaded area indicates extrapolated data). This testing shows a significant decrease in temperature sensitivity by the addition of recycled tire rubber. Similar improvements in temperature sensitivity were previously noted based on viscosity (Figure 1) and penetration (Figure 2) tests.

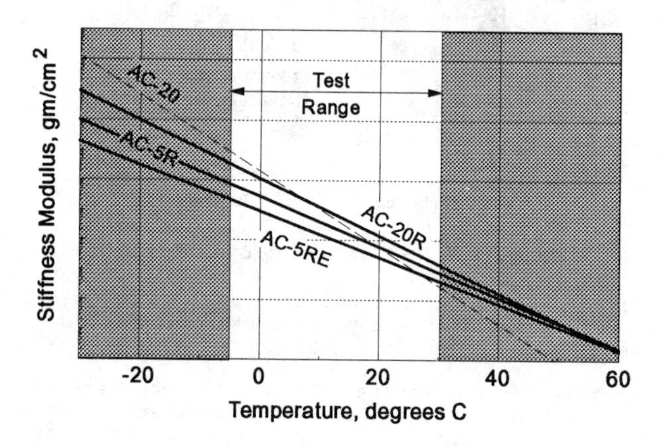

Figure 7 -- Tensile Creep @ 1,000 Seconds

The tensile creep tests also show that similar high temperature stiffness may be achieved with an asphalt-rubber produced with an asphalt cement 2 to 3 grades softer than the neat asphalt cement. The low temperature properties of the asphalt-rubber binder so produced would be much better than the neat asphalt cement. Similar trends may be seen in the softening point (Figure 3) and resilience (Figure 4) test results previously reported.

One fact not shown on the plots is that all the asphalt cements had brittle failures at -5.5°C while the asphalt-rubber remained flexible. This is a further indication of the asphalt-rubbers' improved low temperature properties. An asphalt concrete using an asphalt-rubber binder produced with softer grades of asphalt should result in a mix that is less susceptible to thermal cracking and rutting than a similar asphalt concrete mix produced with a stiffer neat asphalt cement.

MIX DESIGN

Comparison of Mix Design Methods for Asphalt-Rubber Concrete Mixtures

Marshall and Hveem mix design methods were evaluated in Volume 3, "Comparison of Mix Design Methods for Asphalt-Rubber Concrete Mixtures", [4]. Marshall and Hveem mix design procedures were used to determine optimum binder contents using the binders and aggregates previously referenced. The conventional asphalt concrete samples were prepared and tested according ASTM Test Method for Resistance to Plastic Flow of Bituminous Mixtures Using Marshall Apparatus (ASTM D 1559), 50 blows per side, ASTM Test Methods for Deformation and Cohesion of Bituminous Mixtures by Means of Hveem Apparatus (ASTM D 1560) and ASTM Method for Preparation of Bituminous Mixture Test Specimens by Means of California Kneading Compactor (ASTM D 1561). Slight modifications, which are described below, were required for the asphalt-rubber mixes.

Asphalt-rubber Marshall specimens were compacted at 135°C. The samples were allowed to cool overnight before extruding to prevent the specimens from expanding due to the resilient properties of the rubber. An attempt to compact the asphalt-rubber Hveem specimens at 110°C resulted in excessive effective voids. Based on these test results the decision was made to increase the compaction temperature to 149°C.

Some of the conclusions of this research are:

1. Marshall mix design: Rubberized mixtures can be expected to exhibit lower stability and unit weights, and higher VMA and flow than unmodified mixtures; four percent air voids can be obtained with rubberized mixtures. It is recommended that the flow limits be increased; previous suggestions of 22 to 24 for flow appear to be reasonable.
2. Hveem mix design: An increase in compaction temperature from 110 to 149°C produces mixtures that can meet the majority of the traditional Hveem mix design criteria. The Hveem stability limits should be lowered because of the increased lateral deformation per given load that is obtained with the presence of rubber.
3. Comparison of mix design methods: Figure 8 shows recommended binder contents determined by the different design procedures. The asphalt-rubber appears to increase the optimum binder content, regardless of mix design method. Variations of +0.5 percent asphalt were noticed between the two methods, regardless of binders or modifiers.

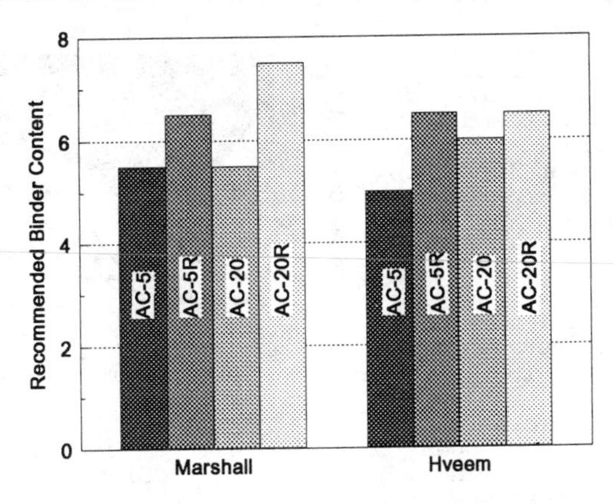

Figure 8 -- Comparison of Mix Design Methods

4. Fundamental material properties: Figure 9 presents resilient modulus vs. temperature for asphalt concrete mixtures produced using four of the binders. A significant reduction in material stiffness at cold temperatures is obtained when asphalt-rubber is added to the mixture. Material stiffness can possibly be increased at warmer temperatures by using asphalt-rubber. The reduced temperature sensitivity was previously noted in binder

tests: viscosity (Figure 1); penetration (Figure 2); and tensile creep (Figure 7).

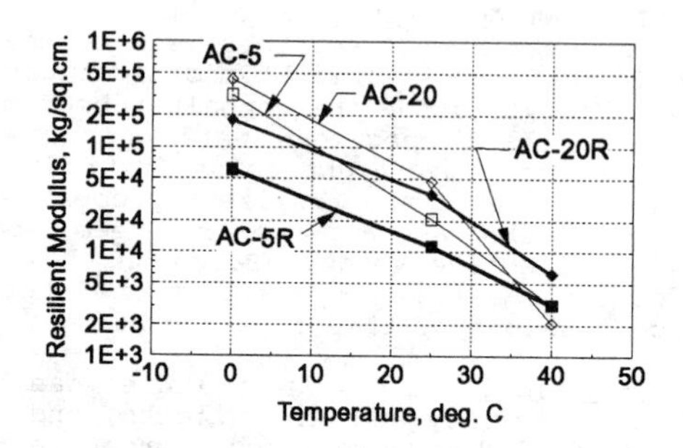

Figure 9 -- Resilient Modulus vs Temp.

Asphalt-Rubber Open-Graded Friction Courses

The use of asphalt-rubber binders in open graded friction courses was evaluated in Volume 8, " Asphalt-Rubber Open-Graded Friction Courses," [5] to: (1) determine the potential benefits of asphalt-rubber binders when used in open graded friction courses; and (2) recommend asphalt cement grades and mix design procedure required to achieve optimum field performance. Mixes were evaluated using:

- Binder Drain Off Tests
- Permeability Tests
- Stripping (Water Sensitivity) Tests

The use of asphalt-rubber binder showed a significant improvement in binder drain off. A comparison of AC-20 and AC-5R is shown in Figure 10. This shows that when asphalt-rubber is used, both the binder content and the mix temperature can be increased. The reduced drain off, even at higher temperatures, is likely due to the higher viscosities of the asphalt-rubber binders at high temperatures as previously shown in Figure 1.

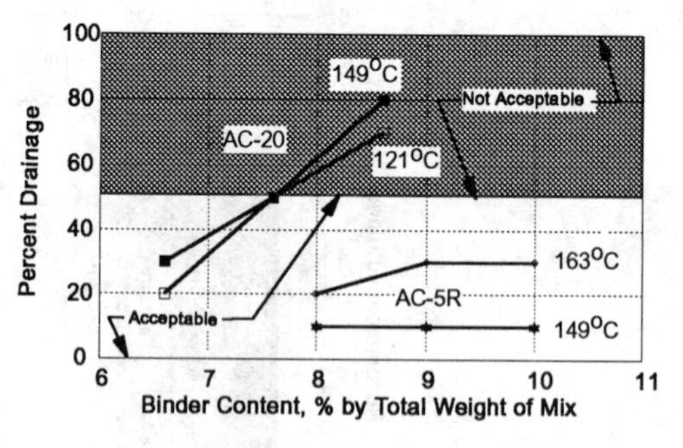

Figure 10 -- Open Graded Mix Drain Off Comparison

Laboratory permeability tests were conducted on open graded mixes produced with each test binder. The test specimens consisted of a 3/4-inch thick open-graded mix on a dense graded mix. The open-graded asphalt cement mixes were mixed at 135°C with binder contents of 6.6, 7.6, and 8.6 percent. The asphalt-rubber samples were mixed at 149°C at binder contents of 8.0, 9.0, and 10.0 percent. Figure 11 presents the results of the permeability tests for four of the mixes tested. This shows equal or better permeability for the asphalt-rubber mixes, even at higher binder contents.

The voids of the compacted open-graded mixes were also evaluated. 15.2 cm (6-inch) diameter, 5.1 cm (2-inch) high specimens were compacted with 25 blows of a Marshall hand compactor on one side. The specimens were weighed in air and water to determine the void content and density. These data showed the asphalt-rubber mixes had higher voids than the asphalt mixes, which agrees with the permeability test results. However, this data also shows increased density for the asphalt-rubber mixes. The combination of higher voids and higher density appears contradictory. Determining the unit weight of a high void mix by weighing the specimens in air and water may have introduced errors due to absorption of water in the mix.

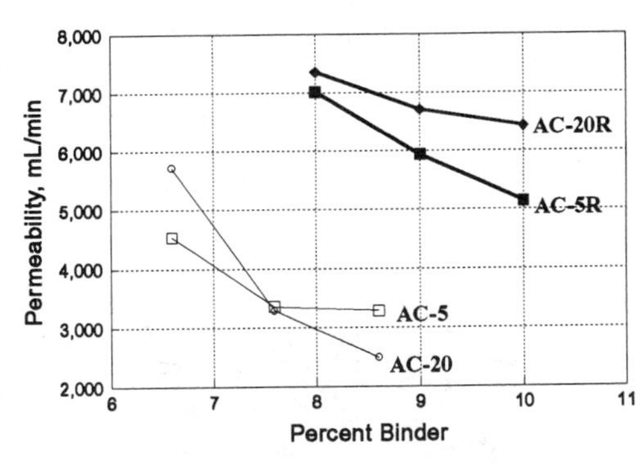

Figure -- 11 Permeability of Open-Graded Mixes

The stripping potential of the mixes was evaluated using the following test procedures:

- ASTM Test Method for Coating Stripping of Bituminous Aggregate Mixtures (ASTM D 1664)
- Texas Boiling Test
- Porewater Pressure Debonding Test

The ASTM procedure is considered the least severe of the tests used and typically identifies only those binders and aggregates with serious stripping problems. All of the binders tested passed the 95% binder retention requirement.

The results of the Texas Boiling and Porewater Pressure Debonding Tests are presented in Figure 12. These tests do show an improvement when asphalt-rubber is used for open-graded mixes. Much of this improvement is likely due to the higher binder contents and resulting increased film thickness.

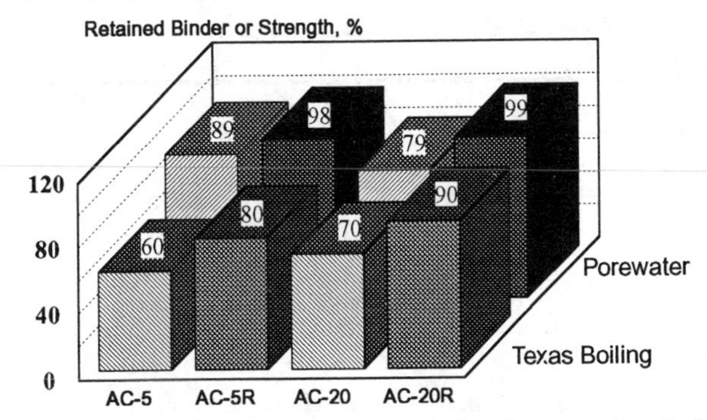

Figure 12 -- Stripping Potential of Open Graded Mix

DENSE-GRADED MIX CHARACTERISTICS

Dense-graded mixes produced using neat asphalt cement and asphalt-rubber binders were tested to compare the following characteristics:

- Permanent Deformation
- Low Temperature Cracking
- Fatigue

Optimum Binder Contents

The binder contents selected for the above testing are shown in Table 1. These binder contents were selected by a committee including the sponsors and the researchers involved. These binder contents were based on mix designs performed by the University of Nevada, Reno (UNR) and U.S. Army Corps of Engineers, Waterways Experiment Station (WES). The binder contents for the unmodified mixes, AC-5 and AC-20, were agreed to at 5.3 and 5.7 percent, respectively. However, the mix designs from UNR and WES for the modified binder did not agree. Therefore a compromise, agreeable to all parties, was made. These binder contents are higher than the recommended binder contents previously reported herein. The researchers involved in evaluating the dense-graded mix characteristics have reported that the binder contents appeared high. It is important that this be considered when evaluating these test results.

Table 1

Type of Binder	Binder Contents Used in Preparing Samples (% by Total Weight of Mix)	UNR Recommend Binder Content (% by Total Weight Of Mix)
AC-5	5.3	
AC-20	5.7	
AC-5RE	8.5	7.7
AC-5R	8.3	7.7
AC-20R	7.9	7.4

Permanent Deformation Characteristics

Permanent deformation characteristics were evaluated in Volume 5, "Permanent Deformation Characteristic of Recycled Tire Rubber Modified and Unmodified Asphalt Concrete Mixtures," [6] using:

- ASTM Proposed Standard Test Method for Unconfined Static Creep Test on Asphalt Mix Specimens (ASTM Sub-committee D04.20)
- Tri-axial, Confined, Repeated Load (SHRP Interim Test Procedure)

Tests were conducted at 25°C and 40°C using both procedures.

Figure 13 shows the results of static creep testing for four of the mixes tested. This testing shows that asphalt-rubber concrete mixtures have reduced permanent deformation at high temperatures when compared to unmodified mixtures.

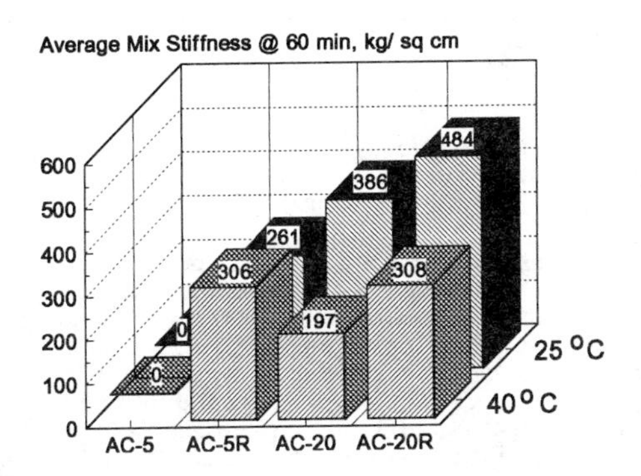

Figure 13 -- Static Creep Test Results

It is interesting that the creep modulus of the AC-5R mix is higher at 70°C than at 25°C. One possible reason for this is the strain at 25°C was almost three times the strain at 70°C.

Figure 14 and 15 present the results of repeated load testing for four of the mixtures. At 25°C this testing shows the AC-20 to be the most resistant to permanent deformation. However, at 40°C the asphalt-rubber mixes have the best performance and the deformation resistance of the AC-20 mixture is considerably reduced. This illustrates that testing at temperatures lower than pavements actually experience in the field may not adequately predict a pavement's resistance to permanent deformation. Thus, permanent deformation testing should be performed at elevated temperatures more representative of field conditions. This conclusion is also supported by both the

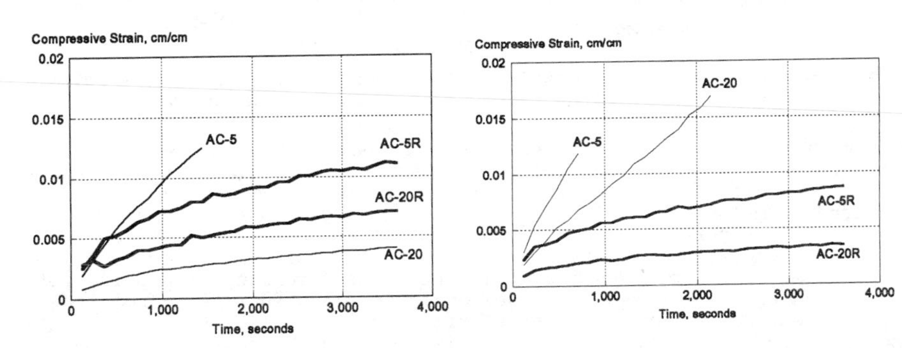

Figure 14 -- Repeated Load Compressive Creep at 25 deg. C

Figure 15 -- Repeated Load Compressive Creep At 40 deg. C

static and dynamic creep data.

Repeated loading should be used for permanent deformation testing. This provides a better model by simulating moving wheel loads and is supported by comparing the static and repeated load tests at 40°C. The static test results indicate only the presence of rubber and nothing about the properties of the base binder. The repeated load testing indicates, in a definite manner, the differences between binders.

Low Temperature Cracking Characteristics

Low temperature cracking characteristics were evaluated in Volume 6, "Low Temperature Cracking Characteristics of Ground Rubber and Unmodified Asphalt Concrete Mixtures," [7] using the following test procedures:

- Indirect Tensile Strength at 1°, -17° and -29°C.
- Constrained Specimen.
- Direct Tension Test at -29°C.

Specimens were also subjected to accelerated aging using NCHRP 9-6(1) AAMAS procedures. Unfortunately all beam specimens used for the constrained specimen test were damaged during the aging and could not be tested. The briquettes used for the indirect tensile strength testing were not damaged during the accelerated aging.

The results of indirect tensile strength tests are presented in Figure 16. It should be noted that the specimens with asphalt-rubber took about twice as long to fail as the unmodified mixes. Since this is a constant strain test, the strains at failure for the asphalt-rubber mixes would be about twice that of the unmodified mixtures. This helps illustrate the conclusion that asphalt-rubber mixtures will exhibit more

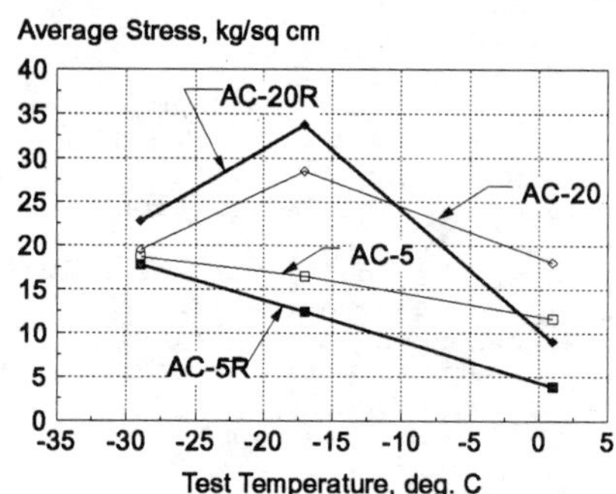

Figure 16 -- Indirect Tensile Strength Test Results

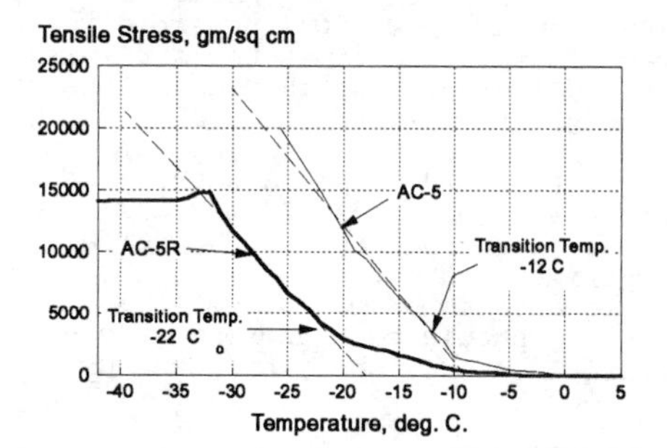

Figure 17 -- Constrained Specimen Test Results

deformation at cold temperatures (i.e. 0°F and -20°F) while maintaining strengths similar to unmodified mixes.

The constrained specimen testing measures the stress required to maintain a specimen at constant length under a constant rate of cooling. Figure 17 shows the results of testing specimens prepared using AC-5 and AC-5R binders. This illustrates that as the temperature drops the stress increases gradually until the "transition temperature" is reached and the stress increases at an accelerated rate. Above the "transition temperature" the mixes still possess viscoelastic properties where the thermal stresses can be relieved through stress relaxation. Below this "transition temperature" the mixture exhibits purely elastic characteristics and the thermal stresses are not relaxed until failure of the specimen.

Figure 17 illustrates the conclusion that the AC-5R binder reduces the transition and fracture temperature by about 10°C (18°F) when compared to the unmodified AC-5 mix.

The AC-20R sample did not show the same improvement using the constrained specimen test compared to the unmodified AC-20 mixture. One possible conclusion is because the rubber particles absorb the light fraction of the asphalt cement, a stiffer base asphalt, such as the AC-20, may be left with only the heavier oils and resins. The resulting mix may be more sensitive to non-homogeneities due to increased stiffness. This leads to the conclusion that softer base asphalts should be used for asphalt-rubber mixtures to resist thermal cracking.

Direct tensile tests were conducted at -29°C using a constant loading rate of 0.025 cm./min.. The measured peak stresses achieved during testing are presented in Table 2. These test results show very little difference between any of the mixtures. The conclusion here is that the extremely low temperature properties of each mix will be similar in as much as they all contain the same supplier of asphalt.

Table 2

Binder Type	AC-5	AC-5R	AC-20	AC-20R
Tensile Strength, kg./sq.cm.	16.38	17.65	16.52	17.65

Little difference was observed in the aging of the AC-5 and AC-5R mixtures. Both mixtures exhibited approximately a 25 percent increase in indirect tensile strength after aging.

Fatigue of Asphalt and Asphalt-Rubber Concrete

The fatigue characteristics of the mixtures were evaluated in Volume 7, "Fatigue of Asphalt and Asphalt-Rubber Concretes," [8] using a deflectometer.

The test equipment applies a repeated central load to a sample about 44.5 cm in diameter. This is a constant stress fatigue test. The stress vs. fatigue plots presented in Volume 7 would indicate the unmodified mixes would have superior fatigue performance at all temperatures. These plots do indicate the mixes with rubber modified binders would have superior fatigue performance at lower stresses only. This data contradicts what was anticipated since all other tests, binder and mixture, showed equal or better performance for the asphalt-rubber mixtures. Also, the higher binder content for the asphalt-rubber mixtures should have, by itself, improved the fatigue resistance. Samples were looked at after testing to evaluate the crack pattern for the type of failure. The crack pattern observed for the asphalt-rubber specimens was not the fatigue pattern observed for the unmodified mixes. This leads to the conclusion that the binder contents of the asphalt-rubber mixes were too high and the constant stress method of fatigue testing used may not be valid for the high binder content asphalt-rubber modified mixtures.

Figure 18 shows the strain vs. repetitions to failure for mixtures with AC-5R and AC-20 binders. These plots were calculated by the author of this summary chapter using equations, moduli and test values presented in Volume 7. The plotted strain values were limited to an equivalent stress value of 21 kg/sq.cm.. This maximum stress value keeps the plot within the stress values tested and the probable maximum tensile strength of the mixes. This plot shows equal or superior performance for the AC-5R mixes. It should be noted the calculated test strains for the asphalt-rubber mixes at a given temperature are significantly greater than for the unmodified mixes. The ability of the unmodified mixes to tolerate the higher strains applied to the asphalt-rubber mixes is unknown. This plot indicates that asphalt-rubber mixes should be tested and evaluated using controlled strain testing rather than controlled stress.

CONCLUSIONS

Asphalt-rubber binders are much less temperature sensitive than the base asphalts they are produced from. This conclusion is supported by viscosity tests, penetration tests at 77°F and 39.2°F and tensile creep data. Asphalt-rubber binders are stiffer at high temperatures and

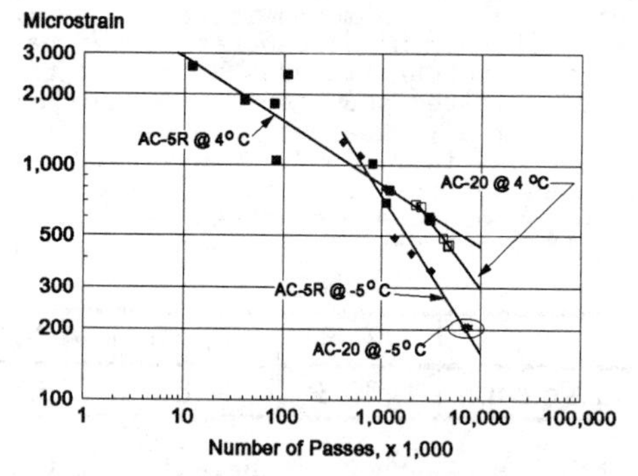

Figure 18 -- Fatigue vs Strain

generally softer at low temperatures than the base asphalt cements.

Asphalt-rubber binders should be produced with softer base asphalts. This is supported by a number of binder tests including softening point, resilience and tensile creep. All of these binder tests show an asphalt-rubber binder produced with an AC-5 asphalt cement to have equal or better high temperature characteristics than an AC-20. Penetration and tensile creep tests show the asphalt-rubber produced with an AC-5 to have superior low temperature properties compared to an AC-20. This conclusion is also supported by permanent deformation and low temperature cracking testing of the dense-graded mixes. This testing shows that an asphalt-rubber dense-graded mix produced with an AC-5 base asphalt is more resistant to both permanent deformation at high temperatures and thermal cracking at low temperatures than a dense-graded mix produced with an unmodified AC-20 or AC-5 asphalt cement.

Asphalt-rubber binders generally showed improved resistance to plant (TFOT) and environmental aging (Weatherometer). Only the viscosity testing of aged binders showed poorer aging for the AC-5RE binder which is produced from an AC-5 base asphalt and extender oil. Penetration testing of the aged binders showed less aging for all asphalt-rubber binders compared to their base asphalt cements.

Asphalt-rubber concrete (ARC) mixes can be designed using Marshall and Hveem procedures with slight modifications. Modifications to the mixing and compaction procedures include: mixing and compaction at higher temperatures than for neat asphalt cements; and allowing the samples to cool before extruding from the molds to prevent volumetric changes in the plugs due to elastic rebound of the rubber.

- For Marshall mix designs, asphalt-rubber mixes can be expected to have lower stability and unit weights, and higher VMA and flow than unmodified mixtures. It is recommended that flow limits for dense-graded mixes be increased to 22 to 24.
- The Hveem stability criteria should be reduced for asphalt-rubber mixes.
- Higher binder contents should be anticipated for asphalt-rubber mixes compared to mixes using neat asphalt cement, regardless of design method.

Testing indicates that asphalt-rubber open-graded friction courses would be more durable, longer lasting and better draining than unmodified open-graded friction courses.

Asphalt-rubber concrete mixes show improved resistance to permanent deformation at high temperatures than unmodified mixes.

Permanent deformation testing should be conducted at high temperature (39°C +) and use repeated loading.

The use of asphalt-rubber in dense-graded mixes shows improved resistance to low temperature cracking compared to unmodified mixes. The use of softer base asphalts (AC-5) for the asphalt-rubber provided the most improvement in low temperature crack resistance.

Fatigue testing of the asphalt-rubber mixes was inconclusive. Further research should be performed on the fatigue response of asphalt-rubber hot mixes.

REFERENCES

[1] Anderton,G., "Physical Properties and Aging Characteristics of Asphalt-Rubber Binders," Construction Productivity Advancement Research Program (CPAR), Corps of Engineers, Vicksburg, MS & Asphalt-Rubber Producers Group, Washington, DC, February 1992.

[2] Hansen,K.R.and Stonex, A. "Tensile Creep Comparison of Asphalt Cement and Asphalt-Rubber Binders," Construction Productivity Advancement Research Program (CPAR), Corps of Engineers, Vicksburg, MS & Asphalt-Rubber Producers Group, Washington, DC, February 1992.

[3] Coetzee, N.F., and Monismith, C.L., "Analytical Study of Minimization of Reflective Cracking in Asphalt Concrete Overlays by use of a Rubber Asphalt Interlayer," Transportation Research Record 700, 1979, pp. 100-108.

[4] Stroup-Gardiner, M., Krutz N. and Epps, J., "Comparison of Mix Design Methods for Asphalt-Rubber Concrete Mixtures," Construction Productivity Advancement Research Program (CPAR), Corps of Engineers, Vicksburg, MS & Asphalt-Rubber Producers Group, Washington, DC, February 1992.

[5] Anderton, G., "Asphalt-Rubber Open-Graded Friction Courses,"Construction Productivity Advancement Research Program (CPAR), Corps of Engineers, Vicksburg, MS & Asphalt-Rubber Producers Group, Washington, DC, February 1992.

[6] Krutz, N.C. and Stroup-Gardiner, M. "Permanent Deformation Characteristics of Recycled Tire Rubber Modified and Unmodified Asphalt Concrete Mixtures" Construction Productivity Advancement Research Program (CPAR), Corps of Engineers, Vicksburg, MS & Asphalt-Rubber Producers Group, Washington, DC, February 1992.

[7] Krutz, N.C. and Stroup-Gardiner, M. "Low Temperature Cracking Characteristics of Ground Rubber and Unmodified Asphalt Concrete Mixtures," Construction Productivity Advancement Research Program (CPAR), Corps of Engineers, Vicksburg, MS & Asphalt-Rubber Producers Group, Washington, DC, February 1992.

[8] Jimenez, R.A.,"Fatigue of Asphalt and Asphalt-Rubber Concretes,"Construction Productivity Advancement Research Program (CPAR), Corps of Engineers, Vicksburg, MS & Asphalt-Rubber Producers Group, Washington, DC, February 1992.

Vince Aurilio[1], Daniel F. Lynch[1], Roger P. Northwood[1]

THE USE AND RECYCLING OF WASTE TIRE RUBBER HOT MIX AT THAMESVILLE ONTARIO

REFERENCE: Aurilio, V., Lynch, D.,F., and Northwood, R., P.,**"The Use and Recycling of Waste Tire Rubber Hot Mix at Thamesville Ontario,"** Use of Waste Materials in Hot-Mix Asphalt, ASTM STP 1193, H. Fred Waller, Ed., American Society for Testing and Materials, Philadelphia, 1993.

ABSTRACT: In support of efforts by the Ontario Ministry of the Environment to dispose of waste tires, the Ontario Ministry of Transportation built a demonstration hot mix pavement incorporating waste tire rubber at Thamesville, Ontario, in 1990 and 1991.

The purpose of the demonstration was to confirm that (a) the construction of asphalt rubber hot mix pavements is environmentally acceptable and complies with worker health and safety regulations; (b) asphalt rubber pavements can be recycled; and (c) to compare the performance and cost of asphalt rubber pavements to standard pavements.

In 1990 the work consisted of placing 6.0 km each of asphalt rubber mix and standard mix. Both mixes were produced in a drum plant. The rubber content of the mix was 2.0% by mass of dry aggregate. In 1991, both mixes were recycled using the same drum plant and a new section of asphalt rubber mix with a finer gradation of rubber was added.

A full schedule of stack testing was carried out in 1990 and 1991 to determine the presence and levels of a wide variety of pollutants. Testing was also carried out to confirm that the process was in compliance with Ontario worker health and safety regulations.

A description of the equipment and the work, data on mix quality, a preliminary performance evaluation, a discussion on the findings of the environmental and worker safety testing and a costing analysis are included in the paper.

KEYWORDS: tires, rubber, Thamesville, hot mix, recycled hot mix, asphalt rubber pavements, demonstration project, pavement performance, mix design, quality assurance, stack emissions, safety

---

In the Fall of 1989, the Ontario Ministry of Transportation (MTO) was approached by the Ontario Ministry of the Environment (MOE) with a proposal to dispose of waste tires in hot mix pavements. This proposal was driven by the mounting problems of tire disposal in Southern Ontario and the commitment made with the introduction of the Ontario Tire Tax.

---

[1]Bituminous Engineer, Section Head and Manager, respectively, Engineering Materials Office, Ontario Ministry of Transportation,1201 Wilson Avenue, Downsview, Ontario M3M 1J8.

Six million passenger tires plus one million truck tires are scrapped each year in Ontario. This translates into about 11 million passenger tire equivalents per annum for disposal. About 10% of the scrap tires are used in industrial applications which leaves about 10 million tires annually going to landfill. Concerns with the potential health and fire hazard problems and the associated problems with disposal of this waste material have led to experimental work directed towards the development of new uses for the scrap rubber derived from these tires.

Rubber from scrap tires had been used in hot mix in Ontario on an experimental basis from time to time. The percentage of fine rubber used in the mix as an aggregate replacement was always 1% and the asphalt cement content was increased by amounts ranging between 0.3% and 0.5%. The purpose of these trials was to improve the performance of the hot mix pavement by reducing/eliminating cracking. This improved performance was not established by any of the trials and in some of the trials the cracking was worse than that found in the standard mix. Since the addition of rubber to hot mix increased the cost of the end-product, the idea always died at that stage. The last of these experiments took place in 1980.

In 1989 the staff of the Ministry of the Environment (MOE) had become aware of a new approach to asphalt rubber hot mix capable of disposing of 3% of rubber in the mix, or four to five tires per tonne of hot mix. This meant that the ten to fifteen million tonnes of hot mix produced annually in Ontario could absorb forty to sixty million scrap tires. Even if all of the hot mix did not contain rubber, there would be more than enough capacity to cope with the annual "production" of scrap tires, plus make quick inroads into the existing tire stockpiles.

Because the aggregate replacement system is new and there is no information available on environmental impacts, worker safety, and long term performance, it was agreed that a major trial of asphalt rubber hot mix should be built to assess the viability and acceptability of this method of tire recycling.

This report evaluates the experience with the use of asphalt rubber hot mix on this project. It discusses the construction and performance aspects of standard and recycled asphalt rubber mixes, as well as the environmental concerns related to air quality and worker safety during construction. Costs of the process are also presented.

OBJECTIVES OF THE TRIAL

The objectives of the trial were:

a) to evaluate the technical aspects of mixing and placing asphalt rubber mix containing 3% of coarse rubber;

b) to mix the material in an oil-fired drum mix plant, since this type of hot mix plant is used on most Ministry work outside of the major urban areas;

c) to obtain exhaust stack air quality measurements for comparison with the regulatory standards;

d) to carry out "Air Quality Assessment" testing for compliance with worker health and safety regulations;

e) to carry out leachate testing in accordance with the regulatory standards;

f) to establish the recyclability of asphalt rubber pavements when using a drum mix plant;

g) to test asphalt rubber mix for compliance with waste disposal regulations;

h) to evaluate the performance of the asphalt rubber mix at high (3%) rubber contents;

i) to determine the viability of using the hot in-place surface recycling process on asphalt rubber pavements, and its compliance with environmental and worker safety regulations; and

j) to obtain reasonably accurate estimates for the cost of asphalt rubber mix.

To meet the above objectives, four test sections were scheduled including a control section with a standard mix. In addition to the asphalt rubber mix, a recycled asphalt rubber mix and a surface recycled mix were proposed. However, it was not possible to carry out these processes in the same construction season and the recycling trials were deferred to 1991.

LOCATION OF THE TRIAL

During the pre-contract preparation stage two suitable locations for the work were identified in South Western Ontario. To encourage competitive bidding the contract was set-up to allow the contractor the option of site selection. Both sites offered pavements that are structurally sound, and carry similar (moderate) traffic volumes which facilitates long term monitoring. The successful bidder - Huron Construction Ltd. of Chatham - chose the Thamesville site on Hwy. 2, 22 km east of Chatham (Fig. 1).

Highway 2 at Thamesville is a two lane rural highway. The terrain is flat and the alignment generally follows the Thames River. The underlying subsoils consist of sand, gravelly loam and clay loam. Traffic volumes are 2050 to 2750 AADT (Annual Average Daily Traffic) and 2250 to 3050 SADT (Summer Average Daily Traffic) with 9% to 10% trucks. The surrounding land use is farming.

The existing pavement was in very good structural condition with a fair ride. The hot mix surface was suffering from slight ravelling in areas of segregation, and there was extensive minor reflection cracking which had been crack filled.

The 7.5 m (24.5 ft.) wide pavement consisted of 130 mm (5 in.) of dense graded hot mix over 230 mm (9 in.) of concrete pavement which had been built in the 1920's. The first hot mix surfacing was placed in 1955 and was resurfaced with recycled hot mix in 1982.

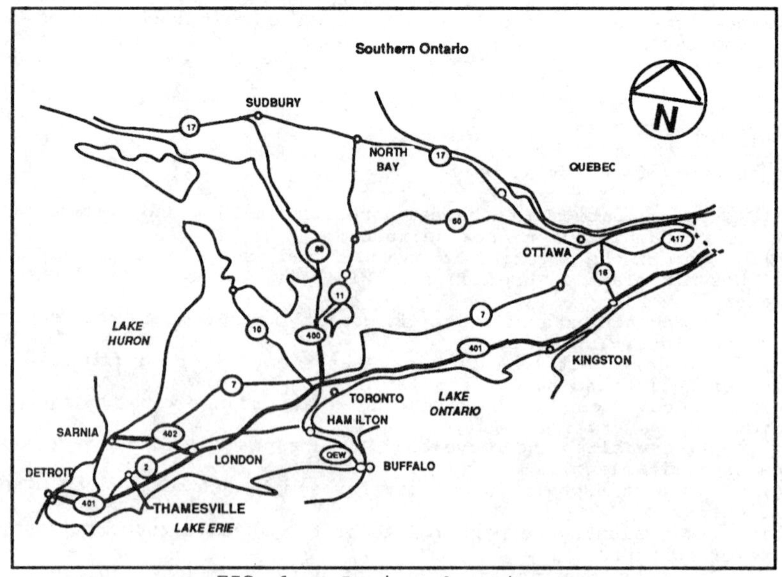

FIG. 1 -- Project location.

LAYOUT OF THE TRIAL

The length of the trial was controlled by three factors:

1) the need to mix and place sufficient standard mix to establish a baseline of environmental and worker safety testing values and provide a control section for future performance evaluation;
2) the need to accommodate three construction processes i.e. asphalt rubber mix, recycled asphalt rubber mix and surface recycling; and
3) sufficient plant operating time for each stack test (6 hours) and for the number of stack tests (three per mix type) required to inspire confidence in the data.

This added up to a total length of 12 km using 6,000 tonnes each of asphalt rubber mix and standard mix. In 1991 2 km sections of each pavement type were to be recycled in a drum plant and surface recycled. The recycling ratio for both recycled mixes was to be 30% reclaimed asphalt pavement (RAP) and 70% new aggregate (30/70).

Because of the concerns with the performance of the 1990 asphalt rubber mix using the No. 4 mesh rubber it was decided to change to No. 10 mesh rubber for the 1991 recycling work. To compare the performance of the 1990 No. 4 mesh rubber with the No.10 mesh rubber an extra trial section approximately 3 km in length was planned with No. 10 mesh rubber. In order to determine if the performance problems with the No.4 mesh rubber used in 1990 were due to non-compliance with the rubber gradation specification a 1 km stretch of asphalt rubber using the No. 4 mesh rubber which complied with the specification was planned for Phase II.

In 1991, the contract for Phase II was awarded by an invitation bid. The work included four test sections, a standard recycled section and a recycled asphalt rubber section each incorporating 30% RAP from the pavement constructed in 1990 and two new sections; one using the original rubber (No. 4 mesh) and one with the No. 10 mesh rubber. Due to recurring problems with the gradation of the No. 4 mesh rubber this section was paved with a standard HL 4 mix.

MIX DESIGN

The proposed standard mix was a minus 19 mm dense graded mix (MTO Designation HL 4 or Asphalt Institute Mix Type IV b). The proposed asphalt rubber mix was HL 4 mix with 3% rubber and an additional 1.5% asphalt cement.

The mix designs for the standard and asphalt rubber mixes for both phases were prepared by three different consultants on behalf of the contractor. In Phase II the design for the recycled hot mix was completed by MTO. The mix designs selected for each phase of the work are summarized in Table 1.

The standard mix initially proposed by the contractor for Phase I incorporated a crushed gravel coarse aggregate and natural fine aggregate. It was found that this mix could not accommodate the 3% rubber requirement. To maximize the percentage of rubber which could be used, the crushed gravel was replaced with a gravel with a higher percentage of crushed material, and 80% of the fine aggregate was replaced with another gravel aggregate of a finer gradation. However, even with these adjustments, the maximum rubber content which could be accommodated in the mix was 2%. The replacement aggregates were also used in the standard mix to maintain uniformity.

In 1991, the same coarse and primary fine aggregates were used in the mixes submitted for Phase II. The contractor however requested permission to use a different fine blending sand.

The mix designs revealed that there are significant differences between the Marshall properties of standard hot mix and asphalt rubber mixes. The Marshall stability of the asphalt rubber mixes was considerably lower than the standard mix. In both phases of the work

TABLE 1 -- Mix Designs.

Job Mix Formula	Phase I		Phase II			
	Standard Mix	Asphalt Rubber Mix	Standard Mix	Recycled Standard Mix	Recycled Rubber Mix	No. 10 mesh Rubber Mix
% Asphalt Cement*	5.3	6.1	5.0	5.0	5.5	6.2
% Coarse Aggregate**	44.5	49.0	39.9	34.5	36.5	50.0
% Fine Aggregate						
(Huron Const. Sand)	45.5	42.0	45.6	35.5	33.5	33.0
(Oxford Sand)	10.0	7.0				
(Komoka Blend Sand)			9.5			15.0
% Reclaimed Asphalt Pavement				30.0	30.0	
% Crumb Rubber	Nil	2.0	Nil	Nil	1.2	2.0
Marshall Properties						
Marshall Stability (N)	10800	5250	8300	10500	6725	5800
Flow (0.25 mm)	10.5	18.8	8.3	7.1	12.5	12.0
Void in Mineral Aggregate (%)	14.0	17.0	15.0	14.2	16.5	18.7
% Air Voids	4.0	4.0	3.4	4.0	3.0	4.0
Bulk Relative Density	2.445	2.337	2.410	2.415	2.365	2.320
Theoretical Max. Density	2.506	2.418	No data	2.513	2.438	2.420

Notes: * By mass of mixture.
** By mass of dry aggregate.

the Marshall stability of the asphalt rubber mixes was below the specified minimum of 6700 Newtons, while the stability of the recycled rubber mix was marginally higher than the minimum value. The testing also indicated that the corresponding flow values were much higher than normal. An increase of about 80% was observed for the mix containing the No. 4 mesh rubber (Phase I) and for the recycled mix. The mix which contained the No. 10 mesh rubber experienced a 45% increase in flow.

The asphalt demand for the asphalt rubber mixes was approximately 1.0% higher than the standard mixes. For the recycled rubber mix the additional asphalt cement required was 0.5%. The voids in mineral aggregate (VMA) of the conventional mixes ranged from 14-15%, which was generally 2-4% lower than the asphalt rubber mixes. Accordingly, the bulk relative densities of the asphalt rubber mixes were consistently lower than the standard mix. Both mixes used in 1990 were designed at an air void content of 4.0%. In 1991 the mixes were designed at 4.0% and 3.0% air voids for the standard recycled hot mix (RHM) and the recycled rubber mix, respectively. The new mix using the No.10 mesh rubber was designed with 4.0% and the standard mix at 3.4% air voids.

CONSTRUCTION

1990 - Phase I

Paving commenced on September 22nd and ended on October 16th with some lost time due to rain. Meteorological data from the MOE station at Courtright, located 55 km northwest of Thamesville, indicated an hourly mean temperature of 17°C for the days when conventional mix was produced (low 8°C and high 26°C) and a mean of 11°C during the asphalt rubber mix production (low 7°C and high 16°C). Wind speeds were recorded at a mean of 14 km/hr for the conventional mix and a mean of 9 km/hr for the asphalt rubber mix.

The hot mix was laid with a Cedarapids Grayhound CR561 rubber tired paver. The compaction train initially consisted of a Bomag vibratory roller (Model BW161AD) for breakdown compaction followed by a static steel wheel Huber roller. A second steel drum Huber roller was used for finish rolling. The mean laydown temperatures were 140°C and 144°C for the conventional and asphalt rubber mixes respectively.

Several problems were observed during compaction which were mainly attributed to the heat retention of the asphalt rubber mix. The temperature after laydown would decrease at a much slower rate than the conventional mix, even in the early mornings when cooler temperatures prevailed. It appears that the cooling rate was not greatly affected by the climatic conditions. As a result, hairline cracks (checking) developed in the mat after intermediate rolling with the steel wheel roller. Roller stop-marks were also noted. In addition the sustained heat in the mat resulted in the migration of the underlying rubberized crack sealer to the surface.

In an attempt to alleviate these problems the contractor delayed the breakdown rolling and allowed the mat to cool prior to secondary compaction. This did not seem to help. A Bros SP 3000 rubber tired roller (as used on the conventional mix) was then used for secondary compaction. The cracking and the roller marks were eliminated, and the rubber tired roller was more effective in sealing the surface of the mat. However, there were some problems with the mix sticking to the wheels of the rubber tired roller. The pick-up material developed a rubbery consistency and consequently could not be compacted back into the mat resulting in surface deficiencies.

1991 - Phase II

Paving began on October 9, 1991 with the recycling of the standard mix; 30% reclaimed asphalt pavement (RAP) was used in the mix. The work was completed in seven days without problems. Recycling of the asphalt rubber mix with the same recycling ratio (30/70) began on October 16, 1991 and concluded on October 19, 1991. The new rubber mix with No. 10 mesh rubber was placed on October 21, 1991.

The weather during Phase II construction was similar to the conditions in 1990. The days were generally sunny and warm with morning ambient air temperatures of $\pm$ 8°C and highs of $\pm$17°C by midday.

In Phase II the same paving equipment was used as in Phase I. The rolling train was the same as developed during the 1990 work. The laydown temperatures averaged 120°C for the standard RHM and varied from 130 to 145°C for the recycled asphalt rubber mix. The temperatures recorded for the No. 10 mesh rubber mix were in the same range as the recycled asphalt rubber mix.
Compaction problems similar to those experienced in 1990 were encountered with both rubber mixes i.e. the mixes were prone to hairchecking which meant that only a few passes were made with the static steel wheel roller, and the mixes were susceptible to severe pickup by the rubber-tired roller. As a result working mat temperatures were about 60°C with the rubber tired roller.

PLANT DESCRIPTION AND PRODUCTION DETAILS

The Contractor produced the hot mix in a Boeing Model 200 parallel flow drum plant. The cold feed system consisted of three bins. The aggregates and RAP entered the drum at the burner end. The burner was manufactured by Genco and was fuelled by #2 stove oil (CGSB CANZ - 3.2 - M89 TYPE Z). The mix was transferred by drag elevator from the drum to a single storage silo.

The plant was equipped with a wet Venturi scrubber system which had 14 spray nozzles with 9.5 mm inlets and 3.2 mm outlets. The system was driven by one electrically powered pump producing 310-345 kPa of pressure in a 50 mm inside diameter pipe. The exhaust stack diameter was 0.84 m and the top of the stack was approximately 12 m above grade.

The production rate for the plant during construction was 140 to 160 tonnes of hot mix per hour which was well within the 150± 20 tph contract requirement. The mean mix temperature at discharge was approximately 154°C. Fuel consumption was 7.43 and 7.95 L per tonne of mix for the conventional and asphalt rubber mixes respectively.

Rubber Handling Equipment

The granulated rubber was delivered to the plant in 22 kg polyethylene bags packaged on pallets. These pallets were lifted up to a platform on top of a 10 m^3 bin. The bags were then opened by two men, and emptied through a grate in the floor of the platform over the storage bin. The rubber was transferred by conveyor to the 5 m^3 storage hopper of the rubber feeder designed and built specifically for this contract. An automatic gate opener was used to control the quantity of rubber filling the 5 m^3 bin. When this bin was full, the gate on the larger bin automatically closed; conversely, when the rubber level was low the gate opened to feed the small bin. A schematic of the rubber handling system is shown on Fig. 2.

The auger of the rubber feeder was driven by a variable speed motor connected to the Master Feed Control in the asphalt plant control van. The plant control panel provided a variable output from 0 to 10 volts and the motor was set at 10 volts direct current. This voltage translated into 560 RPM motor speed. At this output, the rubber flowed at 5 tonnes per hour which was determined by running the auger at this speed and weighing the rubber. The variable drive motor voltage was set to correspond with the required 2% rubber in the mix. As the operating speed (tonnes per hour) increased, the voltage to the auger control increased proportionately. An RPM meter on the control panel was used to verify the proper RPM for variations in the production rate of the plant.

The auger entered the rear of the drum for a distance of 3.5 m and was fixed at the same slope as the drum. The discharge point allowed for approximately one minute of mixing time. The asphalt cement pipe also entered the drum from the rear for a distance of 4.1 m.

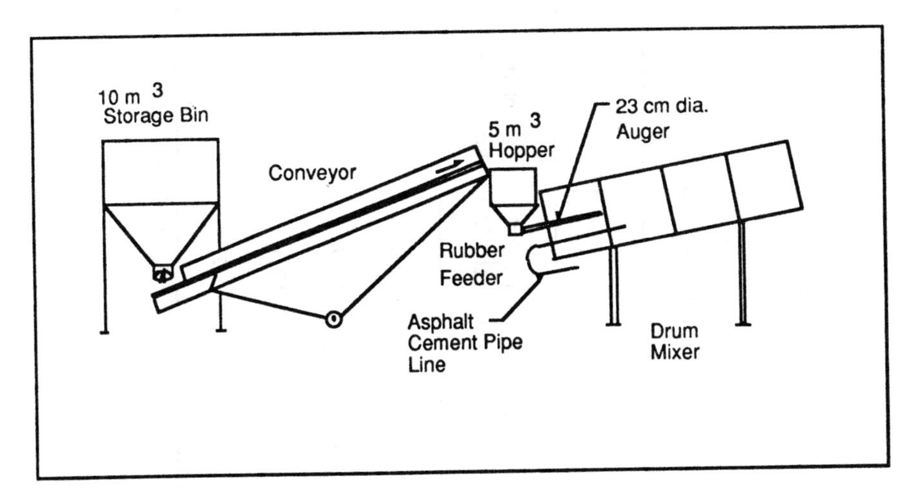

FIG. 2 -- Rubber handling system.

QUALITY ASSURANCE

Acceptance testing for quality assurance for both phases consisted of the routine testing normally conducted on MTO contracts. Random roadway samples were taken during construction for testing to determine asphalt cement content, extracted aggregate gradations, mix properties and recovered penetration in accordance with the contract requirements. Asphalt cement samples were tested for compliance with MTO material specifications and the gradation of the crumb rubber supplied to the contract was tested for compliance with the specified grading. Cores were obtained to verify the compaction achieved in the field.

Marshall Properties

The results (Table 2) indicate that the asphalt rubber mixes exhibit 30-40% lower Marshall stabilities and 28-40% higher flow values than the standard mixes. For the recycled rubber mix the average stability was 12.5% lower and the flow was 15% higher than the standard mix. This is directionally the same as found in the mix designs, however the differences are not as dramatic. It appears that these differences are more pronounced with the 1990 asphalt rubber mix using the No. 4 mesh rubber.

The air voids for both mixes placed in 1990 were lower than expected. The average value for each mix was 2.2% which is considered to be well below the specified minimum of 3.0%. In 1991, the air voids for the standard mix and both recycled mixes did not meet the required specification. The average air voids content for the standard mix was 1.9%; the voids of the recycled mixes were 2.2% (standard RHM) and 2.7% (rubber RHM). The average air void content for the No. 10 mesh asphalt rubber mix was 3.3%.

The Marshall tests were conducted on reheated samples at a

TABLE 2 -- Summary of Marshall test results for production samples.

	Phase I		Phase II			
Marshall Test	Standard Mix	Rubber Mix	Standard Mix	Recycled Standard Mix	Recycled Rubber Mix	No. 10 mesh Rubber Mix
Stability (N)	14514	11016	13018	14088	12523	9266
Flow	12.9	18.1	11.6	11.7	13.4	14.9
VMA (%)	15.3	16.1	13.7	13.1	16.2	18.4
Air Voids (%)	2.2	2.2	1.9	2.2	2.7	3.3

Regional mobile laboratory. Standard briquettes were fabricated using a hand held hammer employing a compaction energy of 75 blows per side.

Asphalt Cement Content and Mixture Gradation

The gradation and asphalt cement content data are shown in Table 3 and Table 4 for Phase I and II, respectively. Extractions were carried out in a field laboratory with the standard Rotarex equipment using a 1, 1, 1, - trichloroethane solvent, and fines correction with the SMM centrifuge in accordance with ASTM Standard Test Methods for Quantitative Extraction of Bitumen From Bituminous Paving Mixtures (D 2172 - 88).

TABLE 3 -- Phase I acceptance test results.

Mix Type	Standard Mix		No. 4 mesh Rubber Mix	
Designated Sieve	Job Mix Formula	Average Gradation	Job Mix Formula	Average Gradation
26.5mm				
19.0 mm	100	100	100	100
16.0 mm	98.3	99.4	98.5	98.1
13.2 mm	91.9	94.5	92.4	91.6
9.5 mm	71.3	78.0	72.0	72.5
4.75 mm	56.0	59.7	51.8	53.0
2.36 mm	47.2	48.7	43.1	42.4
1.18 mm	38.6	38.4	34.8	32.8
600 μm	30.0	29.4	26.6	23.6
300 μm	18.3	19.3	15.8	13.9
150 μm	7.5	8.6	7.0	6.4
75 μm	3.4	4.1	3.2	3.5
% Rubber	-	-	2.00	1.91
% AC	5.30	5.39	6.10	6.16

The test data shows that the mixing process in 1990 was in very good control.

In 1991 the average rubber content was 0.57% and 0.26% higher than required by the Job Mix Formula for the asphalt rubber RHM and the No. 10 mesh rubber mix, respectively and the corresponding asphalt cement contents were also higher by 0.30% and 0.24%. During construction the

TABLE 4 -- Phase II acceptance test results.

Mix Type	Standard Mix		Recycled Hot Mix		Asphalt Rubber Recycled Hot Mix		No.10 mesh Rubber Mix	
Designated Sieve	Job Mix Formula	Average Gradation	Job Mix Formula	Average Gradation	Job Mix Formula	Average Gradation	Job Mix Formula	Average Gradation
26.5mm								
19.0 mm	100	100	100	100	100	100	100	100
16.0 mm	100	99.3	98.9	98.6	99.2	99.3	99.0	98.4
13.2 mm	92.7	87.6	90.0	93.4	91.6	92.7	90.0	91.2
9.5 mm	71.6	66.1	72.1	76.2	72.1	73.2	69.1	71.4
4.75 mm	57.1	54.8	55.0	57.0	53.1	53.7	52.1	53.1
2.36 mm	46.9	44.8	47.8	48.5	45.4	45.3	45.2	45.1
1.18 mm	37.1	34.2	40.8	41.2	39.3	38.0	37.6	37.1
600 µm	25.8	24.4	32.4	32.2	30.1	29.5	27.2	28.0
300 µm	13.1	13.9	19.7	20.2	18.1	17.5	15.1	15.9
150 µm	5.1	5.5	6.5	7.9	7.2	6.1	5.2	5.2
75 µm	2.3	2.9	3.1	4.5	4.2	3.1	3.0	2.4
% Rubber	-	-	-	-	2.00	2.57	2.00	2.26
% AC	5.00	4.95	5.00	5.09	5.50	5.80	6.20	6.44

contractor's process control consisted of extraction tests and bag counts of the rubber used on a daily basis. This process control data indicated that the correct amounts of rubber were added to the mixes.

Gradation of Rubber

The No. 4 mesh rubber supplied to the work in 1990 was coarser than the specification on all but the No. 40 mesh sieve. In 1991 the No. 10 mesh rubber was consistently coarser than specified on the No. 16 mesh sieve even though the specification had been based on information provided by the supplier. The gradation test data for the rubber used in the work is in Table 5.

Pavement Density

The contract specified a minimum of 92% compaction (based on Theoretical Maximum Relative Density) for the lot mean with no single test value less than 90% compaction. For Phase I all samples tested met the requirement. However, there was a significant difference in the overall mean values with the standard mix showing 2.2% better density. In Phase II several lots of the asphalt rubber mixes did not comply with the minimum specified compaction. The standard mix and the standard RHM both met the requirements. Consequently, the overall pavement densities were lower than the standard mixes by 1.9% for the rubber mix (No. 10 mesh) and 2.7% for the asphalt rubber RHM. Areas of asphalt rubber pavement with in-place air voids approaching 10% give rise to some concern with regard to long-term performance. Fig. 3 shows the average compaction of the different mixes based on four sublots per lot.

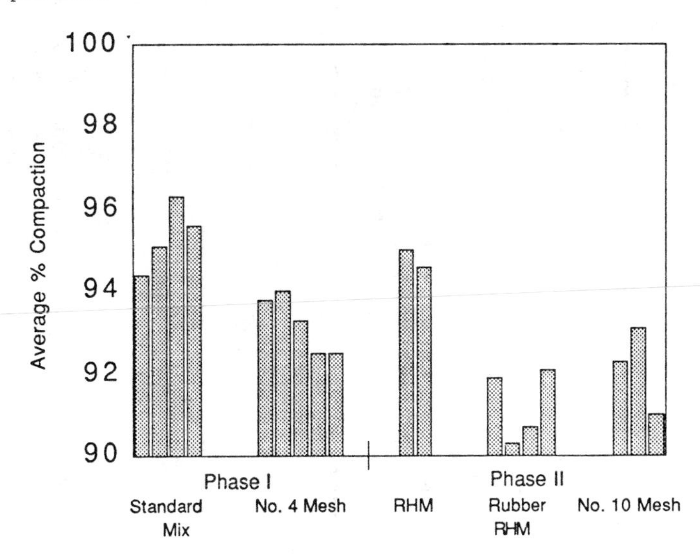

Fig. 3 -- Compaction Test Results by Mix Type

Note: The compaction mean for the single lot of standard mix placed in 1991 (Phase II) was 94.0%.

TABLE 5 - Gradation of rubber used.

No. 4 Mesh Rubber Phase I

Sieve Size	Specification	1	2	3	4	5	6	7	8	Mean
		No. of samples tested								
		Percent Passing								
No.4	100	98.9	99.1	98.8	97.0	99.6	99.9	99.3	99.2	98.8
No.8	70-80	60.1	47.4	67.8	52.0	74.7	50.1	67.4	58.6	59.8
No.10	50-60	45.1	39.2	52.1	41.1	60.1	41.2	58.2	50.1	48.4
No.20	30-40	20.2	18.9	22.1	25.1	26.5	22.2	33.1	31.1	24.9
No.40	5-15	1.6	2.6	10.7	5.6	6.1	4.0	15.4	6.5	6.6

No. 10 Mesh Rubber Phase II

Sieve Size	Specification	1	2	3	4	5	6	7	8	9	10	11	Mean
		Percent Passing											
No.10	95-100	100.0	100.0	100.0	100.0	100.0	100.0	100.0	100.0	100.0	100	100	100
No.16	70-80	54.1	59.1	60.9	60.9	61.5	63.0	65.1	63.8	51.5	54.8	61.1	60.5
No.30	30-40	31.1	32.3	32.5	34.7	34.8	34.9	38.0	37.6	31.4	34.6	34.8	34.2
No.50	10-20	13.6	14.5	14.1	15.0	16.5	15.2	17.5	16.5	12.2	13.5	13.5	14.7
No.200	2-10	1.8	1.5	1.6	1.9	2.1	2.4	2.7	1.3	1.1	1.3	0.6	1.7

Recovered Penetration

Table 6 summarizes the acceptance testing for recovered penetration. The data shows that the recovered penetration values for the asphalt rubber mix produced in 1990 are much higher than the standard mix. The same trend was observed for the 1991 non-recycled mixes, however there is only one test value for the standard mix. It is not clear whether the higher penetration numbers for the asphalt rubber mix are due to the extra film thickness or because some oil was extracted from the rubber. This difference was not experienced with the recycled mixes produced for Phase II.

TABLE 6 -- Recovered penetration test

	Phase I		Phase II		
Lot No.	Standard Mix	Asphalt Rubber Mix	RHM	Rubber RHM	No. 10 Mesh Rubber Mix
1	72	72	nd	63	61
2	74	84	68	78	72
3	74	79	66	70	70
4	68	90		87	66
5	49	83		63	62
Average	67	82	67	72	66

Note: One sample of the standard mix was tested in Phase II ; the recovered penetration was 50 units.

Asphalt Cement

The test results for the 85-100 and 150-200 asphalt cement used in the work are shown in Table 7. The softer grade asphalt cement was used in the recycled mixes. All of the samples tested complied with the specification requirements for Asphalt Cement.

Chemical Analysis

"The contracted chemical analysis specifications for the rubber were met with the exception of the range of measured specific gravity in 1991 [3]." The data for this testing is summarized in Table 8.

TABLE 7 -- Asphalt cement quality.

Asphalt Cement Grade 85/100

Sample No.	Penetration 0.1 mm	Viscosity (mm2/sec)	Ductility (at 4°C)	Flash Unit °C
Phase I				
1	86	334	15+	260+
2	85	324	15+	260+
3	93	314	15+	260+
4	89	319	15+	260+
5	97	318	15+	260+
6	93	302	15+	260+
7	93	236	15+	260+
8	92	310	15+	260+
9	98	316	15+	260+
10	99	283	15+	260+
Phase II				
1	95	349	6+	260+
2	93	344	6+	260+
3	92	345	6+	260+
4	93	348	6+	260+
5	94	340	6+	260+
6	96	351	6+	260+
7	94	349	6+	260+
Specification	85-100	280 min.	6+	232 min.

Asphalt Cement Grade 150/200

Sample No.	Penetration 0.1 mm	Viscosity (mm2/sec)	Ductility (at 4°C)	Flash Unit °C
1	167	225	10+	260+
2	166	236	10+	260+
3	169	225	10+	260+
4	163	254	10+	260+
Specification	150-200	200 min.	10+	218 min.

TABLE 8 -- Summary of chemical properties.

Property	No. 4 Mesh	No. 10 Mesh	Specification
Specific Gravity	1.17	1.15	1.15±0.05
% Carbon Black	30.8	17.8	35 maximum
% Ash	4.5	5.1	8 maximum
% Acetone Extract	15.1	12.3	23 maximum

PAVEMENT PERFORMANCE

Performance surveys

As part of this study the Ministry has initiated a program to evaluate the long term performance of the asphalt rubber pavements on a yearly basis. A consultant was retained in 1990 to carry out a pre-construction evaluation of the existing pavement. The scope of the work included a subsoil investigation, a pavement distress survey, deflection testing and a post construction survey. Some of the original pavement condition data is shown in Table 9. The following year the same firm was awarded a second contract to evaluate the performance of the 1990 work and to extend the pre-construction survey to the new trial sections for Phase II.

In July 1992 Ministry staff visited the site to evaluate the performance of the trials and to carry out an abbreviated crack survey of 500 m of pavement for each trial section. A summary of the data collected is also presented in Table 9.

The data shows that the asphalt rubber mixes have experienced similar transverse cracking to date.

Observations

Field observations were carried out shortly after construction by the consultant to obtain an initial indication of the performance of the trials. Some early problems with coarse aggregate loss and opening of the longitudinal joint along the pavement centerline were identified in the asphalt rubber section.

The site was also visited on numerous occasions by the Ministry's technical staff in 1990 and 1991. At the end of January 1991, both mixes were performing well. However, the following concerns were noted with the asphalt rubber pavement section;

a) some large coarse aggregate particles were dislodged (popouts) from the surface; this seemed to occur wherever a particle of rubber was immediately under the aggregate. The occurrence of this phenomenon was variable and ranged from zero per square metre to 100 per square metre;

b) the longitudinal construction joint at the centerline was beginning to open and some minor ravelling was occurring; and

c) the mat appeared to be more permeable; the rate of drying was slower after rain in comparison with the standard mix.

Reflection cracks were observed in both sections of the pavement constructed in 1990 by Ministry staff and the consultant.

The following observations were noted by Ministry technical staff in July, 1992;

1) the centerline joint for the pavement constructed in 1990 was almost totally open and the popouts in the asphalt rubber section had not changed over the previous year;
2) the centerline joint in the recycled asphalt rubber pavement was 100% open and some minor ravelling was observed along the open joint. The joint opening in the standard recycled pavement was slight and isolated to a few locations;
3) the aggregate loss noted in the recycled asphalt rubber pavement was similar to the 1990 asphalt rubber pavement; and
4) the No. 10 mesh rubber section was performing relatively better than the asphalt rubber pavement built in 1990 for an equivalent service life after one year. The extent of aggregate loss was much less and the centerline joint was slightly open in localized areas.

TABLE 9 -- Comparison of reflection cracking observed to date with the original pavement condition.

Transverse Cracks	Phase I		Phase II		
Type	Standard mix	Asphalt Rubber Mix	RHM	Rubber RHM	No. 10 Rubber Mix
Full Pavement Width	16	36	Nil	32	8
Lane Width	45	55	11	14	
Partial	129	73	148	124	
Total	190	164	159	170	21
Original Pavement	267	225	254	253	No Data
% Cracks Reflected	71	73	63	67	-

ENVIRONMENTAL AND WORKER SAFETY TESTING

The testing was conducted jointly by Ortech International in 1990 and Air Testing Services in 1991 under contract to Dames and Moore, Canada and by the Ontario Ministry of the Environment's Mobile Air Pollution Sampling (MAPS) mobile laboratory.

The testing program included monitoring of stack emissions and wastewater generated at the plant from the production of the conventional and asphalt rubber mixes and air quality sampling in the vicinity of workers at the plant and on site during paving. The following discussion is based on the information provided by Dames & Moore, Canada in a summary report prepared for the MTO. The data included in this report was tabulated by Dames & Moore, Canada.

Preliminary assessment results of this work to date indicate that there are some detectable differences in air emissions between standard hot mix production and asphalt rubber mix production. The production of RHM resulted in higher air emissions and exposure levels. The worst case impact of emissions from the production of both recycled mixes exceeded Ontario air quality criteria. Particulate emission levels increased during asphalt rubber mix versus conventional hot mix production and increased with RHM production [3].

Air Emissions

The air emission components tested are broadly classified as follows:

- inorganic components;
- combustion gases;
- chlorinated compounds;
- polyaromatic hydrocarbons; and
- volatile organic compounds.

Inorganic components include particulate material and metals. Combustion gases include carbon monoxide, carbon dioxide, nitrogen oxides, oxygen, total hydrocarbons, total reduced sulphur, hydrogen bromide, hydrogen chloride, hydrogen fluoride, nitric acid, nitrous acid, phosphoric acid and sulphuric acid. The chlorinated compounds analyzed were dioxins, furans, polychlorinated biphenyls, chlorobenzenes and chlorophenols.

Four tests were carried out on the standard mix and three tests on the asphalt rubber mix: the fourth test scheduled for the asphalt rubber mix was abandoned because of rain.

Sampling with the Metals Sampling Train and the Trace Organics Sampling Train was conducted isokinetically on the main exhaust stack at a height of approximately 7.8 m downstream from the exhaust fan which complied with the guidelines of at least eight stack diameters downstream and at least two stack diameters upstream of a flow disturbance.

Sampling with the Fluorides Sampling Train was conducted isokinetically on the main exhaust stack at a distance of approximately 6.1 m downstream from the exhaust fan. This was not an ideal sampling location so the number of sampling points was increased in accordance with the Source Testing Code.

Sampling with the Volatile Organics Sampling Train (VOST) was similarly conducted at a distance of 6.1 m downstream from the exhaust fan. Sampling of combustion gases was conducted at a distance of approximately 4.6 m downstream from the exhaust fan.

In 1990 the levels of particulate matter exceeded air quality standards. These levels were greater for the asphalt rubber mix. In 1991, the use of a finer mesh rubber did not result in greater particulate emissions. The production of RHM generally resulted in an increase of particulate emissions. However the asphalt rubber RHM did not exhibit a significant increase over the standard RHM. The levels of iron in particulate emissions did not exceed the standards in 1990 but did exceed the standards in 1991.

The measured levels of benzene emitted did not increase with the addition of rubber to hot mix. However, in 1990 this project released 1.5 times the total benzene emission commonly allowed per year by the MOE. The 1991 project released nearly eight times this limit. "To date, the MOE has not, made a determination on what an allowable emission level would be for this case [3]." A comparison of the worst case values for emission rates and the standards is presented on Table 10.

Physical Testing Results

Samples of the wastewater generated by the air emissions control equipment at the plant were analyzed to determine if on-site disposal would be permissible under current water quality regulations. The 1990 testing indicated that there were numerous exceedances of the standards. This experience was duplicated in the 1991 program. Table 11 outlines a summary of water testing data. Based on this testing the water quality was not suitable for disposal in storm sewers. The level of manganese was found to be much higher than the allowable limit for discharge into the sanitary sewer system .

Samples of asphalt pavement, rubberized asphalt pavement, rubber and the sludge generated by the air emissions control equipment were tested for leachable metals and benzo-a-pyrene. No exceedances were found [3]. The test results indicate that these materials can be disposed of in normal sanitary landfills.

Worker Safety

The air samples obtained during hot mix production at the plant and at the paving site were tested for a variety of metals, particulates, selected volatile organic compounds and heavier organic compounds. Sampling was carried out in a fixed location within the worker's breathing zone and on the workers with an air pump connected to a filter (for capturing particulates) and/or a contaminant absorbing tube for capturing gases or vapor.

The only exposure problems of concern in 1990 were with the measured levels of total particulates during the production of the asphalt rubber mix. One test for particulate was 1.6 times the established maximum level. The 1991 testing did not produce any exceedances of the total particulates however, levels of organics exposure did increase during the production of the recycled hot mix.

"The testing for selected volatile organic compounds, coal tar pitch volatiles and polyaromatic hydrocarbons did not exceed the Occupational Standards in 1990 or 1991 [3]." The maximum exposure levels measured are outlined in Table 12.

The conclusions and recommendations made in the Dames & Moore, Canada summary report are as follows:

1) The production and lay down of hot mix and more specifically asphalt rubber hot mix has the potential for air emission and occupational airborne contaminant exposure problems.
2) The use of natural gas as firing fuel for the plant process may decrease both particulate and benzene emissions levels. Strict process control and in depth maintenance of emissions control equipment can also help to reduce maximum emission rates.
3) The use of more advanced pollution control equipment than currently available at this facility may be required in order to attain full air emissions regulatory compliance for both conventional and rubberized asphalt production.
4) Proper disposal of process water should be practiced in conjunction with plant operations.

TABLE 10 -- Worst case values combined with emission rates for 1990 and 1991 compared to standards using MOE air dispersion models.

	MAXIMUM EMISSION RATE g/sec		MAXIMUM CONCENTRATION AT A GROUND LEVEL POI 1990				MAXIMUM CONCENTRATION AT POI IDENTIFIED IN CERTIFICATE of APPROVAL APPLICATION 1990			
COMPOUND	HL4	RUBBER ASPHALT	ug/m3* HL4	% OF STD	ug/m3* RUBBER ASPHALT	% OF STD	ug/m3* HL4	% OF STD	ug/m3* RUBBER ASPHALT	% OF STD
SUSPENDED PARTICULATE	1.92	5.49	133	133	381	381	142	142	405	405
PARTICULATE CONCENTRATION IN STACK GAS (g/Rm3)	0.32	0.91		138		395				
IRON	0.061985	0.034857	4.30	43	2.42	24	4.58	46	2.57	26
CONTINUOUSLY ANALYZED 1/2 HOUR AVERAGE MAXIMUMS										
NITROGEN OXIDES	1.180	0.780	82	16.4	54	10.8	87	17.4	58	11.5
SULPHUR DIOXIDE	0.390	0.390	27	3.3	27	3.3	29	3.5	29	3.5
CARBON MONOXIDE	16.43	3.97	1141	19.0	276	4.6	1213	20.2	293	4.9
TOTAL REDUCED SULPHUR	0.090	0.040	6.2	15.6	2.8	6.9	6.6	16.6	3.0	7.4
TOTAL DIOXINS ng/sec	22.40	9.10	1.56***	0.04	0.63***	0.14	1.65***	0.37	0.67***	0.15
TOTAL FURANS ng/sec	401.80	3.10	2790***		0.22***		2967***		0.23***	
DIOXINS AND FURANS			0.065*****	6.5	0.002*****	0.2	0.070*****	7.0	0.002*****	0.20
TOTAL PCB's mg/sec	0.005072	0.002284	0.35***	0.078	0.16***	0.035	0.37***	0.083	0.17***	0.037
BENZO-A-PYRENE mg/sec	0.007674	0.014691	0.53***	16	1.02***	31	0.57***	17	1.08**	33
TOLUENE	0.027665	0.015299	1.92	0.096	1.06	0.053	2.04	0.102	1.13	0.056
ETHYLBENZENE	0.015760	0.002692	1.09	0.027	0.19	0.0047	1.16	0.029	0.20	0.0050
XYLENES	0.092928	0.022801	6.45	0.28	1.58	0.069	6.86	0.30	1.68	0.073
STYRENE	0.009995	0.007458	0.69	0.17	0.52	0.13	0.74	0.18	0.55	0.14
BENZENE	0.009556	0.008526	1324****	166	1182****	148				

TABLE 10 -- Continued

COMPOUND	MAXIMUM EMISSION RATE g/sec 1991			MAXIMUM CONCENTRATION AT A GROUND LEVEL POI 1991					
	ug/m3* HL4 RECYCLE	RUBBER ASPHALT RECYCLE	RUBBER ASPHALT	ug/m3* HL4 RECYCLE	% OF STD	ug/m3* RUBBER ASPHALT RECYCLE	% OF STD	ug/m3* RUBBER ASPHALT	% OF STD
SUSPENDED PARTICULATE	10.06	9.41	0.97	804	804	752	752	78	78
PARTICULATE CONCENTRATION IN STACK GAS (g/Rm3)	1.52	1.39	0.14		659		604		60
IRON	1.033	0.258	0.0147	82.59	826	20.63	206	1.18	12
CONTINUOUSLY ANALYZED 1/2 HOUR AVERAGE MAXIMUMS									
NITROGEN OXIDES	0.539	0.706	0.679	43	8.6	56	11.3	54	10.9
SULPHUR DIOXIDE	0.896	0.893	0.264	72	8.6	71	8.6	21	2.5
CARBON MONOXIDE	5.82	7.47	0.54	465	7.8	597	10.0	43	0.7
TOTAL REDUCED SULPHUR	0.080	0.103	0.028	6.4	16.0	8.2	20.6	2.2	5.6
TOTAL DIOXINS ng/sec	9.99	31.4	2.79	0.80***	0.18	2.51***	0.56	0.22***	0.05
TOTAL FURANS ng/sec	22.5	64.1	4.44	180***		5.12***		0.35***	
DIOXINS AND FURANS				0.006*****	0.58	0.017*****	1.70	0.001*****	0.13
TOTAL PCB's mg/sec	nd	nd	nd						
BENZO-A-PYRENE mg/sec	0.0156	0.0252	nd	1.25**	38	2.01**	61		
TOLUENE	0.541600	0.797680	1.200620	43.3	2.2	63.8	3.2	96.0	4.8
ETHYLBENZENE	0.049647	0.029698	0.087236	3.97	0.099	2.37	0.059	6.97	0.174
XYLENES	0.047847	0.035267	0.111040	3.82	0.17	2.82	0.12	8.88	0.39
STYRENE	0.006533	0.009464	0.015962	0.52	0.13	0.76	0.19	1.28	0.32
BENZENE	0.027544	0.034764	0.002681	2776****	347	3254****	407	212****	27

TABLE 10 -- Continued

COMPOUND	MAXIMUM CONCENTRATION AT POI IDENTIFIED IN CERTIFICATE of APPROVAL APPLICATION 1991							
	ug/m3* HL4 RECYCLE	% OF STD	ug/m3* RUBBER ASPHALT RECYCLE	% OF STD	ug/m3* RUBBER ASPHALT	% OF STD	ONTARIO 0.5 hr. POI "STANDARD"*	
SUSPENDED PARTICULATE	856	856	801	801	83	83	100	FOR d<44um
PARTICULATE CONCENTRATION IN STACK GAS (g/Rm3)							0.2	g/Rm3 ONT. REG. 469/87
IRON	87.90	879	21.95	220	1.25	13	10	
CONTINUOUSLY ANALYZED 1/2 HOUR AVERAGE MAXIMUMS								
NITROGEN OXIDES	46	9.2	60	12.0	58	11.6	500	
SULPHUR DIOXIDE	76	9.2	76	9.2	22	2.7	830	
CARBON MONOXIDE	495	8.2	636	10.6	46	0.8	6000	
TOTAL REDUCED SULPHUR	6.8	17.0	8.8	21.9	2.4	6.0	40	
TOTAL DIOXINS ng/sec	0.85***	0.19	2.67***	0.59	0.24***	0.05	450	***
TOTAL FURANS ng/sec	1.91***		5.45***		0.38***			
DIOXINS AND FURANS	0.006*****	0.61	0.018*****	1.81	0.001*****	0.14	1	ratio calculation
TOTAL PCB's mg/sec								ng/m3
BENZO-A-PYRENE mg/sec	1.33**	40	2.14**	65			3.3	ng/m3
TOLUENE	46.1	2.3	67.9	3.4	102.2	5.1	2000	
ETHYLBENZENE	4.22	0.106	2.53	0.063	7.42	0.186	4000	
XYLENES	4.07	0.18	3.00	0.13	9.45	0.41	2300	
STYRENE	0.56	0.14	0.81	0.20	1.36	0.34	400	
BENZENE							800	grams total per year

NOTES:
- POI - Point of Impingement
- "STANDARD" - Ontario Ministry Of the Environment Standard, Tentative Standard Guideline, Provisional Guideline (May 1990 List)
- Concentrations at POI's derived using worst case from Ontario Ministry of the Environment Regulation 308 (of the Environmental Protection Act) Dispersion Model's
- nd - not detected

* = ug/m3 - except as noted
** = ng/m3
*** = pg/m3
**** = grams emitted by this project
***** = ratio calculation
STD = STANDARD

- blanks - are a result of
a) no data available
b) no POI concentration calculated due to nondetectable anlaysis for compound
c) no applicable standard identified for comparison purposes

TABLE 11 -- 1990 and 1991 summary of detected water analysis results compared with "Standards"

Parameter	Units	CCME	PWQO	Model Sewer Use By-law	1990 Maximum	1991 Effluent Maximum	1991 Lagoon Maximum
General Chemistry							
pH		F,G 6.5 to 9.0	6.5 to 8.5	5.5 to 9.5		7.39	8.15
Field pH		F,G 6.5 to 9.0	6.5 to 8.5	5.5 to 9.5		7.72	
Biochemical Oxygen Demand	mg/L	*	*	300		232	1
Total Suspended Solids	mg/L	*	*	350		61006	9
Field Dissolved Oxygen	ppm	F,G 5 to 9.5	4 to 8	*		8.20	
Total oil and grease	mg/L	*	***	165	21	334	182
Mineral oil and grease	mg/L	*	***	15		148	106
Animal oil & grease	mg/L	*	***	150		185	75
Fluoride	mg/L	I 1.000	*	10		0.8	0.2
Chloride	mg/L	I 100-700	*	1500		100.9	28.1
Sulphate	mg/L	F,G 1000	*	1500		2195	482
Phenols	mg/L	F,G 0.001	0.02	1		3.80	<0.0005
Cyanide (FREE)	mg/L	F,G 0.005	0.005	2		0.030	<0.0001
Sulphide	mg/L	*	*	INOFFENSIVE		0.117	<0.002
Total Kjeldahl Nitrogen	mg/L	*	*	100	31	26	<2
Total Phosphorous	mg/L	*	G 0.02	*		2.787	0.007
Volatile Suspended Solids	mg/L	*	*	*		70910	8

TABLE 11 -- Continued.

Parameter	Units	CCME	PWQO	Model Sewer Use By-law	1990 Maximum	1991 Effluent Maximum	1991 Lagoon Maximum
Metals							
Silver	mg/L	F,G 0.0001	0.0001	5		<0.005	<0.005
Aluminum	mg/L	F,G 0.005-0.1	G 0.075	50	1.69	63.00	1.29
Arsenic	mg/L	F,G 0.05	0.1	1		<0.05	<0.05
Barium	mg/L	DW 1.0	*	*	0.179	0.987	0.035
Bismuth	mg/L	*	*	5		<0.1	<0.1
Cadmium	mg/L	F,G 0.0002-0.0018	P,G 0.00015	1	ND	<0.005	<0.005
Cobalt	mg/L	I 0.05	P,G 0.0004	5	ND	0.16	<0.05
Chromium	mg/L	F,G 0.002 - 0.02	0.10	5	ND	0.08	<0.01
Copper	mg/L	F,G 0.002 -0.004	P,G 0.0001	3	0.02	0.38	<0.01
Iron	mg/L	F,G 0.3	0.3	50	4.14	170.86	0.28
Lead	mg/L	F,G 0.001 - 0.007	0.00005	5	ND	0.48	0.06
Manganese	mg/L	I 0.2	*	5	6.79	13.80	0.02
Molybdenum	mg/L	I 0.01-0.05	P,G 0.01	5	0.30	<0.2	<0.2
Nickel	mg/L	F,G 0.025 - 0.15	0.03	3	ND	0.21	<0.05
Phosphorous	mg/L	*	*	10	ND	3.7	<0.5
Antimony	mg/L	*	P,G 0.007	5		0.1	<0.1
Selenium	mg/L	F,G 0.001	0.1	5		<0.1	<0.1
Tin	mg/L	*	*	5		<0.2	<0.2
Titanium	mg/L	*	*	5	0.038	1.610	0.029
Vanadium	mg/L	I 0.1	P,G 0.007	5	0.006	0.151	<0.005
Zinc	mg/L	F,G 0.03	0.02	3	0.06	2.04	<0.01

TABLE 11 -- Continued.

Parameter	Units	CCME	PWQO	Model Sewer Use By-law	1990 Maximum	1991 Effluent Maximum	1991 Lagoon Maximum
Organics							
Chlorobenzene	μg/L	*	15	*		0.2	ND
Benzene	μg/L	F,G 300	P,G 100	*	2.70	17.1	ND
Ethylbenzene	μg/L	F,G 700	P,G 8	*	0.20	3.5	ND
Styrene	μg/L	*	P,G 4	*		0.4	ND
Toluene	μg/L	F 300	P,G 0.8	*	0.70	4.2	ND
O-Xylene	μg/L	DW 300	P,G 40	*	0.30	0.7	ND
M-Xylene + P-Xylene	μg/L	DW 300	P,G (2-M,30-P)	*	0.90	1.0	ND
1-Methylnaphthalene	μg/L	*	P,G 1	*		15.2	ND
2-Methylnaphthalene	μg/L	*	P,G 2	*		20.6	ND
Acenaphthene	μg/L	*	*	*		5.0	ND
Di-n-octyl phthalate	μg/L	*	DO 0.2	*		ND	10.6
2,4-Dimethylphenol	μg/L	*	10.5	*		47.5	ND
4-Nitrophenol	μg/L	*	P,G 48	*		344	ND
m-Cresol & p-cresol	μg/L	*	P,G (1-M,1-P)	*		643	ND
o-Cresol	μg/L	*	P,G 1	*		175	ND
Phenol	μg/L	*	P,G 5	*	96	2050	ND
1,2,4-Trichlorobenzene	μg/L	*	0.5	*		0.010	ND
Pentachlorobenzene	μg/L	*	0.003	*		0.159	ND

TABLE 11 -- Continued.

NOTES:

BLANK	No Data Available
CCME	Canadian Council of Ministers of the Environment, Interim Canadian Environmental Quality Criteria for Contaminated Sites, September, 1991
PWQOs	Ontario Provincial Water Quality Objectives and Guidelines, July, 1991
ppm	Parts per Million
*	No Standard
***	Field Observation Parameter (Identify the Presence of Visible Film) not Conducted Due to Foaming of Process Water
<	None Detected Above the Following Detection Limit
DO	PWQO Is For Other phthalates
DW	Canadian Drinking Water Quality Guidelines Aesthetic Objective
F	Canadian Fresh Water/Aquatic Life Criteria
Field	Field Measurement of Parameter
G	Guideline
I	Irrigation
L	Livestock Watering
ND	Not Detected
P	Proposed

TABLE 12 -- Continued.

Contaminant	Time Weighted Average Exposure Level mg/m3	1990 Maximum Exposure Level Measured mg/m3	1991 Maximum Exposure Level mg/m3	Mode	Sample Location
Coal Tar Pitch Volatiles "or"	0.2	0.13	0.45	one	plant worker
Asphalt Fume for comparison	5		1.17	three	rubber shoveller
purposes			2.22	two	screed man
			1.68	one	paver driver
Benzo(a)pyrene	2*, ***	NA	<0.0001		
Chyrsene	2*, ***	NA	<0.0001		
Pyrene	**	NA	0.00085	three	screed man
Fluorene	**		0.0006	three	screed man
Phenanthrene	**		0.0008	three	screed man
Anthracene	**		0.0002	three	screed man

NOTES:

** For these three parameters, there are no reported exposure level standards
NA Not Analyzed
2* Exposure to be avoided (2 mg/m^2)
*** American Conference of Governmental Industrial Hygenists (ACGIH) Threshold Limit Value (TLV)
MODE ONE HL4 Recycle Production and Laydown
MODE TWO Rubberized Recycle Prodution and Laydown
MODE THREE Rubberized (No. 10 Mesh) Virgin Production and Laydown

TABLE 12 -- Comparison of highest airborne worker exposure levels to regulatory guidelines 1990 and 1991 summary results.

Contaminant	Time Weighted Average Exposure Level mg/m^3	1990 Maximum Exposure Level Measured mg/m^3	1991 Maximum Exposure Level		
			mg/m^3	Mode	Sample Location
Benzene	16	<0.1	<0.036		
Toluene	376	0.1	19.84	one	paver driver
Ethyl Benzene	435	<0.1	0.05	one	paver driver
Xylenes (sum of isomers)	435	<0.1	0.23	one	paver driver
Styrene	213	<0.2	0.74	one	plant worker
Napthalene	52	<0.3	NA		
Total Particulate	10***	15.9	6.04	three	rubber shoveller
Aluminum, water soluble	2	0.047	0.008	three	plant worker
Barium	0.5	0.0013	0.00266	three	rubber shoveller
Chromium II and III total Cr measured	0.5	0.0016	0.00015	one	screed man
Cobalt	0.05	<0.0007	<0.00009		
Copper	1	0.0019	0.00036	two	rubber shoveller
Iron, water soluble	1	0.0737	0.0106	three	plant worker
Lead	0.15	0.0023	0.0013	two	rubber shoveller
Magnesium as MgO	10	0.052	0.0504	three	plant worker
Nickel, acid digestate	1	<0.001	0.00083	two	paver driver
Titanium, as TiO_2	10	0.0251	0.0012	two	rubber shoveller
Vanadium, as V_2O_5 respirable	0.05	0.0438	0.00024	three	rubber shoveller
Zinc as ZnO	10	0.3738	0.029	three	rubber shoveller

COSTS

The additional materials and handling costs for the asphalt rubber mix at Thamesville were -

Phase I:	Rubber 2% @ $370.00/t	= $7.40
	Asphalt Cement @ $200.00/t @ 0.8%	= $1.60
	Handling	= $1.40
	Aggregate upgrading	= $2.60
		$13.00/t
Phase II:	Rubber 2% @ $814.00/t	= $16.30
	Asphalt Cement @ $200.00/t @ 1.2%	= $2.40
	Handling	= $1.40
	Aggregate upgrading	= $2.60
		$22.70/t
Phase II: (RHM)	Rubber 1.2% @ $814.00/t	= $9.75
	Asphalt Cement @ $200.00/t @ 0.5%	= $1.00
	Handling	= $1.40
	Aggregate upgrading	= $2.60
		$14.75/t

For non-experimental work the typical price for HL 4 mix in 1990 and 1991 in Southern Ontario was $35.00 per tonne. This means that asphalt rubber mix containing No. 4 mesh rubber is approximately 37% more expensive than standard mix and asphalt rubber mix containing No. 10 mesh rubber is 65% more expensive than standard mix. This does not include the cost of about $100,000.00 in modifications to the hot mix plant as it is not reasonable to attach all of these costs to this work alone.

FUTURE WORK

The emission testing has identified some problems with the production of asphalt rubber mixes in drum plants and the implications of these findings are presently unresolved from an environmental perspective. At this time the hot in-place surface recycling has been delayed pending the development of an air emission sampling protocol.

SUMMARY OF FINDINGS

The following summary is based on the evaluation of the test results and the performance of the trial sections to date along with the environmental and worker safety testing conducted for both phases of the work.

1) The asphalt rubber mixes were more difficult to compact in the field;

2) The control and the asphalt rubber pavement sections have experienced similar reflection cracking to date; the aggregate loss (popouts) observed was more severe in the rubber mixes containing the coarser crumb rubber used in 1990;

3) It is expected that the asphalt rubber mixes will exhibit a reduction in service life due to the lower degree of compaction obtained;

4) The No. 10 mesh rubber mixed placed in 1991 appears to be performing better than the No. 4 mesh rubber mix placed in 1990.

5) The cost of the asphalt rubber mixes produced in both years was higher than the standard mixes;

6) The total particulate and benzene emissions from both phases of

this project exceeded the established standards;

7) The leachate testing indicates that waste sludge collected at the plant and the asphalt mixes (standard and rubber) can be disposed of in normal sanitary landfills; the water analysis shows that the wastewater generated on-site is not suitable for discharge into storm drainage facilities;

8) In 1990 one test for total particulate exceeded the established maximum level for worker safety. The standard levels were not exceeded in 1991. Levels of worker exposure to organic contaminants did not exceed regulatory limits, and the asphalt fume standard was not exceeded in 1990 or 1991.

REFERENCES

[1] Lawrence, C. E., Killackey, B. J., and Lynch, D. F., "Experimental Hot Mix Pavement with Scrap Tire Rubber at Thamesville, Ontario, Report #1", Canadian Technical Asphalt Association, Proceedings - Thirty-Six Annual Conference, November, 1991.

[2] Lynch, D. F., and Northwood, R. P., "The Waste Tire Demonstration Project at Thamesville, Ontario", Transportation Association of Canada 1991 Annual Conference.

[3] Dewit, M. and Pefhany, C. M., "Summary of Draft Report on Environmental and Health and Safety Issues Relating to the Scrap Tires in Bituminous Pavement Project in 1991", Prepared by Dames & Moore, Canada, for the Ontario Ministry of Transportation, July, 1992.

Glass

Randy C. West[1], Gale C. Page[2], and Kenneth H. Murphy[3]

EVALUATION OF CRUSHED GLASS IN ASPHALT PAVING MIXTURES

REFERENCE: West, R. C., Page G. C., and Murphy K. H., "**Evaluation of Crushed Glass in Asphalt Paving Mixtures**," Use of Waste Materials in Hot-Mix Asphalt, ASTM STP 1193, H. Fred Waller, Ed., American Society for Testing and Materials, Philadelphia, 1993.

ABSTRACT: Laboratory tests conducted to determine the effects of substituting fifteen percent crushed glass for a portion of the fine aggregate in an asphalt paving mixture indicate that the mixtures containing either coarse or fine crushed glass had lower Marshall stabilities and dry tensile strengths compared to a control mixture. The percentage of retained tensile strengths (TSR's) indicated that the mixture containing fine crushed glass was prone to moisture damage. The mixture containing coarser crushed glass compared favorably with the control mixture with respect to TSR. The liquid antistripping agent was apparently ineffective in mitigating stripping in the tensile strength tests. Conversely, boil tests with only the glass indicated that the coarse glass stripped more than the fine glass, and that some improvement was affected by the antistripping additive.

KEYWORDS: glass, glasphalt, waste materials, stripping, aggregates

In 1988 the Florida Legislature passed Senate Bill 1192 that directed the Florida Department of Transportation (FDOT) to expand, where feasible, its use of recovered waste materials in highway construction. Of the solid waste stream, one component which has been identified for potential use in highway construction is glass [1].

Waste glass has been proposed for use as an aggregate replacement or supplement in highway embankments, unbound aggregate bases, portland cement concrete, and asphaltic concrete. Glass is produced primarily from silica sand, which may be used in each of these applications. Waste glass has also been used to make glass beads for reflective highway paints, and to make glass fiber for fiber-reinforced concrete.

Glass can also be recycled in the manufacture of glass containers and other glass products, but it must be free of contaminants and sorted by color [2]. The cost of separating and cleaning waste glass from municipal-type waste streams, however, is economically prohibitive. Glass cleaned and separated at the source (i.e. household separation) has proven to be economical and is successful enough to eliminate

[1]Bituminous Research Engineer, Florida Department of Transportation, P.O. Box 1029, Gainesville, FL 32601.

[2]State Bituminous Materials and Research Engineer, Florida Department of Transportation, P.O. Box 1029, Gainesville, FL 32601.

[3]State Bituminous Engineer, Florida Department of Transportation, P.O. Box 1029, Gainesville, FL 32601.

landfilling the majority of glass waste.

Reprocessing waste glass into aggregate size particles appears to require relatively little effort. Sorting glass into colors is not necessary, and minor amounts of contaminants can be handled without difficulty. Although it is essential that the glass be crushed to a uniform and consistent gradation, this can be achieved with a simple crushing operation.

In addition to technical feasibility, economic feasibility of utilizing waste glass as an aggregate replacement must also be considered. In short, the cost of the crushed glass including the avoided cost of landfilling, must be competitive with use of a conventional aggregate.

PURPOSE AND SCOPE

The purpose of this study is to evaluate the feasibility of using crushed glass as a replacement for a portion of the fine aggregate in asphalt concrete mixtures. Laboratory experiments were conducted to determine the effects that a small percentage of crushed glass would have on standard mix properties. The economics of using crushed glass as an aggregate substitute in Florida is also briefly discussed.

DISCUSSION

The use of glass as an aggregate in asphalt concrete mixtures was first investigated in the early 1970's in the United States and in Canada. Some of the salient findings from previous research efforts are briefly reviewed in this section.

Stripping

The potential of moisture damage of asphalt concrete mixtures containing glass has been a significant concern of most research efforts. It is well accepted that asphalt does not adhere well to glass particles due to their smooth surface, and since glass is hydrophilic, the asphalt coating is easily displaced by water. Severe stripping of glass-aggregate bituminous mixtures has been reported when antistripping additives were not used [3]. However, the use of hydrated lime as an antistripping additive has been successful in reducing the stripping problem [4,5].

Decreased Cooling Rate

Several demonstration projects using "glasphalt" have noted that the glasphalt mat cooled at a slower rate than a conventional mat during construction [5,6]. This has been attributed to the efficiency of heating the glass particles due to the shorter conduction path of the thinner particles, and also to the insulating nature of the glass particle orientation in the mat [6]. The decreased cooling rate has been cited as an advantage due to the extended time for placement and compaction in cool weather paving [6].

Glass Degradation

Degradation, or breaking, of glass particles in mixtures during construction has been documented in a number of reports [3,4,7]. Large and elongated glass particles often fracture during handling, mixing, and compaction. A maximum size of 3/8 inch has been recommended for glass particles in asphaltic mixtures [4].

Skid Resistance

Field constructed surface mixes containing waste glass have been reported to have adequate skid resistance, and in at least one case has demonstrated improved wet skidding resistance and reduced surface wear compared to a conventional mix [8]. However, ravelling of glass particles from the surface was a problem on another project [3].

Performance

Many of the initial reports on test sections constructed with glass in asphalt mixtures have indicated good performance [7,8,9,10]. However, there is very little information available on long-term performance.

Cost

The cost of waste glass in most cases exceeds the cost of conventional aggregates [4]. Prices for color-sorted waste glass have recently been driven upward by increases in demand from the glass recycling market. Currently, the value of color sorted glass ranges from $20 to $60 per ton. Commingled waste glass, on the other hand, has a negative value and is most often landfilled at a cost of up to $50 per ton. Costs for handling, crushing and hauling commingled waste glass to be suitable for use as an aggregate will vary.

LABORATORY STUDY

Materials

Three asphalt concrete mixtures were tested to determine the effects of crushed glass:

Control Mix--A Type S-III (structural) mixture with all limestone aggregates as typical of the Miami area (Table 1).

TABLE 1--Gradation of the control mixture.

Material:	limestone	limestone	screenings	JMF[1]
Percentage:	15	35	50	
Sieve	Percentage Passing Sieve			
½"(13 mm)	100	100	100	100
⅜"(10 mm)	99	100	100	100
#4	15	75	100	79
#10	5	5	85	45
#40	4	4	42	23
#80	3	3	17	10
#200	2.0	2.0	4.2	3.1

[1]JMF=Job Mix Formula

Coarse glass mixture--Same as the control mixture except that the limestone screenings were reduced to thirty-five percent and fifteen percent coarse crushed glass was added (Table 2).

Fine glass mixture--Same as the control mixture except that the limestone screenings were reduced to thirty-five percent and fifteen percent with fine crushed glass was added (Table 3).

The three mixtures were prepared with eight percent AC-30 for Marshall testing. Specimens were also prepared for indirect tensile strength testing with AC-30 and with AC-30 containing 0.5 percent liquid antistripping additive.

The crushed glass was obtained from the New York City Department of Sanitation through the RRT Empire Company. This glass was of assorted colors and contained some minor amounts of non-glass matter such as paper, metal and plastic from lids. This glass was used "as received" for the coarse glass mixture. It was further crushed to obtain the fine glass gradation. The gradation of the fine crushed glass, which meets the requirements for Local Materials, Section 902 of the FDOT specifications, more closely approximates the gradation of the screenings than does the coarse glass.

TABLE 2--Gradation of the mixture with coarse glass.

Material:	limestone	limestone	screenings	glass	JMF[1]
Percentage:	15	35	35	15	
Sieve	Percentage Passing Sieve				
½" (13 mm)	100	100	100	100	100
⅜" (10 mm)	99	100	100	97	99
#4	15	75	100	58	72
#10	5	5	85	14	34
#40	4	4	42	1	17
#80	3	3	17	1	8
#200	2.0	2.0	4.2	0.4	2.5

[1]JMF=Job Mix Formula

TABLE 3--Gradation of the mixture with fine glass.

Material:	limestone	limestone	screenings	glass	JMF[1]
Percentage:	15	35	35	15	
Sieve	Percentage Passing Sieve				
½" (13 mm)	100	100	100	100	100
⅜" (10 mm)	99	100	100	100	100
#4	15	75	100	100	79
#10	5	5	85	85	45
#40	4	4	42	53	25
#80	3	3	17	20	10
#200	2.0	2.0	4.2	8.1	3.7

[1]JMF=Job Mix Formula

Marshall Properties

Three specimens for each mixture were prepared and tested for Marshall properties in accordance with ASTM Test Method for Resistance to Plastic Flow of Bituminous Mixtures Using Marshall Apparatus (D 1559). Maximum specific gravity of each mixture was determined in accordance with ASTM Test Method for Theoretical Maximum Specific

Gravity and Density of Bituminous Paving Mixtures (D 2041), and bulk specific gravity of each specimen was measured in accordance with ASTM Test Method for Bulk Specific Gravity and Density of Compacted Bituminous Mixtures Using Saturated Surface-Dry Specimens.

Results indicate that both the coarse and fine glass caused a reduction in stability (Table 4). Average Marshall stability of the coarse glass mixture was 15 percent lower than the control mixture, and average stability of the fine glass mixture was 12 percent lower than the control mixture.

TABLE 4--Marshall properties.

	Bulk Density g/cm^3 (pcf)	Air Voids %	Marshall Stability kN (lbs)	Marshall Flow mm (.01 in)
Control mix	2.173 (135.6)	3.8	11.1 (2470)	2.8 (11)
Coarse glass mix	2.191 (136.7)	2.9	9.3 (2090)	3.6 (14)
Fine glass mix	2.155 (134.5)	4.1	9.6 (2163)	2.8 (11)

Note: Values reported are averages of three tests.

Tensile Strengths

For each mixture, six Marshall type specimens (three specimens with and three without antistripping additive) were also prepared for indirect tensile strength testing to evaluate stripping potential according to ASTM Test Method for Effect of Moisture on Asphalt Paving Mixtures (D 4867).

Tensile strengths of all glass mixtures (conditioned and unconditioned, with and without antistripping additive) were less than the respective tensile strengths of the control mixtures (Tables 5 & 6). For the mixtures without antistripping additive, the percent retained tensile strength, or TSR's, were: 70% for the control mixture, 85% for the coarse glass mixture, and 52% for the fine glass mixture. Similar results were determined for the mixtures containing antistripping additive.

These results indicate that the coarse glass mixture was affected less by the moisture conditioning than was the control mixture. The fine glass mixture, however, appears to be quite susceptible to moisture damage.

Several factors may have accounted for the significant difference between the coarse glass mixture and the fine glass mixture. Though both mixtures contained the same amount of glass by weight, the fine glass mixture contained a significantly larger area of glass surface due to the finer glass gradation. This increased the number of weak interface bonds between asphalt and glass surfaces where moisture could interfere and cause a loss of strength. The greater surface area also resulted in a thinner asphalt film thickness which further increased the moisture susceptibility of the glass-asphalt interface. In contrast, the coarse glass mixture had a coarser gradation than both the fine glass mixture and the control mixture which resulted in a thicker asphalt film thickness and may be the reason for coarse glass mixture's improved resistance to the effects of moisture conditioning. Another factor which may have affected the strength of the glass-asphalt bond was the age of the crushed glass faces. The fine glass was freshly crushed and not allowed to weather prior to the laboratory testing. This has been shown to adversely affect the stripping potential of some aggregates.

TABLE 5--Tensile strengths, mixes without antistripping additive.

	Bulk Density g/cm³(pcf)	Air Voids %	UCS[1] kPa (psi)	CS[2] Kpa (psi)	TSR[3] %
Control mix	2.167 (135.2)	4.0	1082 (157)	758 (110)	70
Coarse glass mix	2.184 (136.3)	3.0	855 (124)	724 (105)	85
Fine glass mix	2.159 (134.7)	4.0	834 (121)	434 (63)	52

Note: Values reported are averages of thee tests.
[1]UCS=Unconditioned tensile strength, [2]CS=Conditioned tensile strength,
[3]TSR=Tensile strength ratio.

TABLE 6--Tensile strengths, mixes with antistripping additive.

	Bulk Density g/cm³(pcf)	Air Voids %	UCS[1] kPa (psi)	CS[2] kPa (psi)	TSR[3] %
Control mix	2.170 (135.4)	4.0	1069 (155)	814 (118)	76
Coarse glass mix	2.188 (136.5)	3.0	855 (124)	738 (107)	86
Fine glass mix	2.159 (134.7)	4.0	945 (137)	421 (61)	45

Note: Values reported are averages of three tests.
[1]UCS=Unconditioned tensile strength, [2]CS=Conditioned tensile strength,
[3]TSR=Tensile strength ratio.

The results of the tensile strength tests also show that the antistripping additive used in this study was not effective in improving the aggregate-asphalt or glass-asphalt bond for any of the mixes.

Boil Tests

A visual evaluation of the potential for stripping of the glass particles was made after subjecting the two glass gradations to a simple boil test. Samples of 100 percent glass were prepared with 1000 grams of both glass gradations and eight percent AC-30 with and without antistripping additive. Each sample was placed in tap water and boiled vigorously for ten minutes, then the water was decanted and the samples were air dried in a flat pan. The dried samples were transferred to a pyrex pie dish and placed on a light table to visually evaluate the amount of stripped glass particles.

The estimates of the percent of glass particles which were stripped of asphalt were:

Coarse glass, AC-30 without antistripping additive	25%
Coarse glass, AC-30 with antistripping additive	10%
Fine glass, AC-30 without antistripping additive	15%
Fine glass, AC-30 with antistripping additive	5%

These results contrast with the indirect tensile strength results in two ways. First, in the boil test, the coarse glass appeared stripped to a greater degree than the fine glass, and second, the antistripping additive improved the glass-asphalt bond by about 10 percent for both glass gradations.

ECONOMIC ANALYSIS

In Florida, a significant number of municipalities and communities have been successful with voluntary curbside recycling programs. Newspaper, metal cans, and plastic and glass bottles are recycled in most cases. The glass bottles are separated by color during collection. Sorted glass in Florida now sells from $30 to $55 per ton.

A minor amount of waste glass also exists as broken cullets of commingled (mixed colors) glass. As such, these stockpiles of waste glass are not suitable for recycling and therefore have no market value. Disposal of this glass in landfills would cost as much as $33 per ton.

In order for this type of waste glass to be economically feasible for use in asphalt mixes, the total cost (including processing and hauling) must be competitive with local material (i.e. non-commercial sand).

In Florida, local materials are typically mined from pits at or near the asphalt plant site. Since these materials require very little handling and no processing, their costs are often negligible compared to the other mix components. In southeast Florida, however, there are few suitable deposits of such local materials and consequently fine aggregates are obtained commercially. The cost of commercial fine aggregate in the Miami area is $4.00/ton with a haul cost of $.15/ton/mile. Therefore, for the use of glass to be economically feasible, the total cost of the glass, including hauling and crushing, must also be near the same figures.

It is expected that in the southeast Florida metropolitan area where the solid waste disposal problem is acute and where a considerable amount of glass waste is generated, waste glass may be an economic alternative to commercial fine aggregate.

CONCLUSIONS

Based on the results of this limited laboratory evaluation it is concluded that:

1) The standard mix design procedure used by the FDOT (i.e. Marshall and voids analysis) may not adequately assess the durability of asphalt mixtures containing glass and should be supplemented with moisture conditioning and tensile strength testing.

2) The stability of Marshall compacted specimens decreased by more than 12 percent when 15 percent of the fine aggregates are replaced with either coarse or fine glass.

3) The dry indirect tensile strengths of Marshall specimens decreased by more than 20 percent when 15 percent crushed glass (fine and coarse) was substituted for the a portion of the fine aggregate.

4) Moisture conditioning of Marshall specimens caused a 15 percent decrease in tensile strength of the coarse mixture, and a 50 percent decrease in tensile strength of the fine glass mixture. In comparison, the control mixture decreased in tensile strength by 30 percent when subjected to the moisture conditioning procedure.

5) Visual observations of boil test samples indicate that improved bonding of the asphalt binder to the glass particles is achieved when 0.5 percent antistripping additive is used. TSR's of the laboratory prepared mixtures, however, indicate that the antistripping additive was ineffective in reducing moisture damage.

6) It is unlikely that the use of crushed glass in asphalt mixtures

will be economically feasible where suitable local materials are available at or near the asphalt plant.

7) The use of crushed glass in asphalt mixtures may be economically feasible in areas where suitable fine aggregates can only be obtained commercially and the asphalt plant is near the glass source.

RECOMMENDATIONS

The use of crushed glass in asphalt mixtures should be approached with caution and consideration given to economic feasibility and the possible sacrifice in mixture performance. The following technical restrictions are recommended:

1) The amount of crushed glass in the asphalt mixture shall not exceed 15 percent (by weight of the total aggregates).

2) The crushed glass shall be processed to have 100 percent passing the 3/8 inch sieve and no more than 8 percent passing the No. 200 sieve.

3) In addition to meeting the conventional mix design properties as required in the FDOT Standard Specifications, the mixture shall also have a tensile strength ratio equal to or greater than 75% when subjected to ASTM Test Method for Effect of Moisture on Asphalt Paving Mixtures (D 4867).

4) Asphalt mixtures containing crushed glass shall contain an antistripping additive which can be demonstrated to satisfactorily improve the moisture damage resistance of the mixture.

5) Crushed glass shall not be used in either dense-graded (FC-1 & FC-4) or open-graded (FC-2) friction course mixtures.

It is recommended that Special Provisions, to include the above restrictions, be developed for use by FDOT and local governments. These Special Provisions could be included in specific contracts involving asphalt paving where a source of crushed glass is available. However, use of the glass should be optional to the contractor to allow the most economical materials to be used.

REFERENCES

[1] Thompson, Paul Y., The Utilization of Certain Waste By-Products for Highway Application, Executive Summary submitted to the Florida Department of Transportation, University of Florida, August 1989.

[2] Murphy, Robert J., Research Requirements for the Recycled and Reuse of Solid Waste Materials, Florida Highway Technology and Industry Council, February 1989.

[3] Malisch, W. R., Day, D. E., Wixson, B. G., and Anderson, K. O., Use of Domestic Waste Glass as Aggregate in Bituminous Concrete, Highway Research Record No. 307, Highway Research Board, 1970.

[4] Hughes, Chuck S., Feasibility of Using Recycled Glass in Asphalt, Virginia Transportation Research Council, March 1990.

[5] Sussman, W. A., Reclaimed Glass Aggregate Asphalt Pavement, Highway Focus, Vol. 8, No. 2, April 1976.

[6] Dickson, P. F., Cold Weather Paving with Glasphalt, Symposium on Secondary Uses of Waste Glass, January 1973.

[7] Pavement is Half Glass and Concrete Waste, Engineering News Record, Vol. 189, No. 17, 1972.

[8] Peleg, M., Study of Road Friction, Technion-Israel Institute Technology, December 1972.

[9] Malisch, W. R., Day, D.E. and Wixson, B. G., Glasphalt: New Paving Material Completes First Canadian Trial, Engineering and Contract Record, November 1970.

[10] Malisch, W. R., Day, D. E., and Wixson, B. G., Final Report: Use of Domestic Glass for Urban Paving, Missouri University, Rolla, 1975.

Ash

Robert W. Styron,[1] Frederick H. Gustin,[1] and Terry L. Viness[2]

MSW ASH AGGREGATE FOR USE IN ASPHALT CONCRETE

REFERENCE: Styron, R. W., Gustin, F. H., and Viness, T. L., **"MSW Ash Aggregate for use in Asphalt Concrete,"** Use of Waste Materials in Hot-Mix Asphalt, ASTM STP 1193, H. Fred Waller, Ed., American Society for Testing and Materials, Philadelphia, 1993.

ABSTRACT: The concept of taking a throw-away material and converting it into an aggregate that meets construction industry standards would benefit both the environment and the construction industry. Municipal Services Corporation (MSC) has developed a synthetic aggregate manufactured from ash recovered from the incineration of municipal solid waste (MSW).

In the patented MSC process, the ash is broken down into its basic components, mixed with binders and chemical fixation agents to permanently immobilize heavy metals of concern, and pelletized on a pan pelletizer into a graded aggregate. In development of the product, special attention was given to the absorption of asphalt cement which was initially high, but which was brought to within acceptable limits through a process modification.

The aggregate, known under the trade name "TAP", or Treated Ash Product, was shown in the laboratory to meet Minnesota DOT physical testing requirements for aggregates in bituminous mixtures. In preliminary asphaltic concrete mix design testing, Marshall Stability properties were in the range of 2000 - 2500 lb (8900 - 11,120 N), but asphalt

---

[1] Director of Product Development and Senior Project Engineer, formerly with Municipal Services Corporation, 1000 Cobb Place Blvd, Bldg. 400, Kennesaw, Georgia 30144, now with JTM Industries, Inc. and KBK Enterprises, Inc., respectively, at the same address.

[2] Materials Engineer, Law Engineering, 396 Plasters Avenue, N.E., Atlanta, Georgia 30324.

demand was high (7.5%). The problem was addressed by coating the TAP with lime. Subsequent Marshall Stability values averaged 2540 lb (11,300 N) at an asphalt cement content of 5.7%. In environmental testing, the TAP leachate met current federal drinking water standards after being ground to a fine powder and subjected to the U.S. EPA's TCLP (SW 846-1311), an acid leaching test.

KEYWORDS: MSW ash, synthetic aggregate, TAP, chemical fixation, waste materials, incineration, waste-to-energy, asphalt concrete, Marshall Stability, leachability, heavy metals, TCLP.

INTRODUCTION

The disposal of solid waste is a matter of increasing concern for municipalities and state governments throughout the United States. In 1990, 195.7 million tons of municipal solid waste (MSW) were generated in the United States, or 4.3 pounds per person per day. After materials recovery for recycling and composting, 3.6 pounds per person per day were discarded. Without additional source reduction, the amount of waste generated in 1995 is expected to reach 208 million tons [1].

One method of reducing this massive volume is through incineration. Incineration reduces the volume of MSW by as much as 90 percent, and its weight by 75 percent. In 1992, there were 190 MSW incineration plants in operation in the U.S., with a total annual capacity of 35.5 million tons [2]. The majority of these plants are modern waste-to-energy facilities which can produce up to 40 MW of electricity for sale to local electric utilities, and which are required to meet increasingly stringent air pollution standards with each new permit. The most common type of plant is the "mass-burn" plant, which requires little if any waste preparation, other than screening out certain oversize or otherwise undersirable materials prior to combustion. The other major type of plant burns "refuse-derived fuel" (RDF), which entails a considerable amount of material recovery and waste preparation before combustion.

In spite of the acknowledged benefits of burning MSW to produce energy, there remain large quantities of ash residues that must be managed either through disposal, or some form of recovery or utilization. Problems arise in utilization if the solubility levels of heavy metals, such as lead and cadmium, exceed the federal toxicity limits for hazardous wastes when subjected to leachability testing.

In most cases, fly ash is collected in a baghouse or electrostatic precipitator, and then combined with quenched bottom ash. Large chunks of ferrous metals are removed using rotating drum magnets, and the combined bottom and fly ash is transported to the disposal facility in a moisture conditioned, dust-free condition.

The basic physical nature of the combined bottom and fly ash is that it is exceedingly heterogeneous. It consists of agglomerated fine ash and scrubber lime particles interspersed with and coating larger pieces of metals, glass, rock, slag, and small quantities of uncombusted materials such as paper. The principal minerals in the finer particles of ash are identified by x-ray diffraction analysis as kaolinite, smectite, quartz, and carbonate. Concentrations of the major chemcial components, as shown in Table 1, are similar to those of coal ash or even many soils.

TABLE 1 -- Typical total composition analysis of metals reported as oxides in MSW mass burn combined bottom & fly ash[3]

Silica (SiO_2)	40 - 50%
Alumina (Al_2O_3)	5 - 15%
Ferric Oxide (Fe_2O_3)	12 - 25%
Calcium Oxide (CaO)	8 - 15%
Sodium Oxide (Na_2O)	3 - 6%
Magnesium Oxide (MgO)	1 - 2%
Potassium Oxide (K_2O)	0.75 - 1.5%
Titanium Dioxide (TiO_2)	0.75 - 1.5%
Sulfur Trioxide (SO_3)	0.50 - 1.5%
Phosphorus Pentoxide (P_2O_5)	0.50 - 0.75%
Cupric Oxide (CuO)	0.06 - 0.15%
Lead Oxide (PbO)	0.04 - 0.22%
Zinc Oxide (ZnO)	0.12 - 0.22%
Loss on Ignition @ 750°C	1 - 3%

[3] Analysis performed by Commercial Testing and Engineering, Golden, Colorado.

ASH-TO-AGGREGATE PRODUCTION PROCESS

Municipal Services Corporation (MSC) has developed and patented a process for the production of a construction-grade aggregate known by the trade name of "TAP", or Treated Ash Product. The MSC ash-to-aggregate production process (see Figure 1) consists of the removal of ferrous and non-ferrous metals and any unprocessible materials, particle size reduction, chemical fixation of the ash residue, addition of binders, pan pelletization, and curing. Testing has been conducted to qualify the TAP synthetic aggregate for use in bituminous pavement and in portland cement concrete using tests normally conducted on natural mineral aggregates.

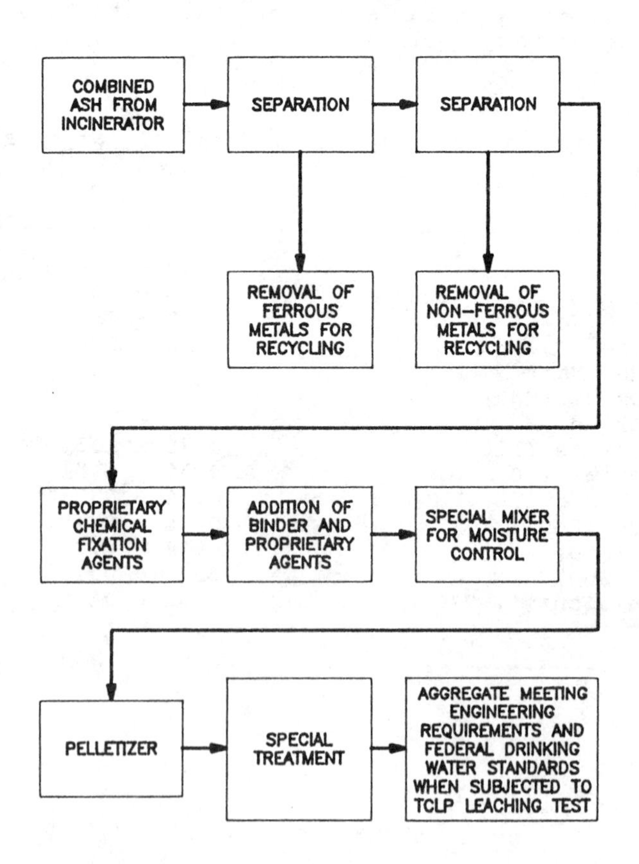

FIG. 1 -- MSC TAP Production Process

The ash-to-aggregate, or TAP, production process has been scaled up to a pilot plant at the MSC/JTM Industries (JTM) Materials Testing and Research Facility, located near Atlanta, GA. The facility can process a variety of combustion by-products into materials of construction.

ENVIRONMENTAL TESTING

The heterogeneous nature of municipal solid waste and its combustion by-products results in an ash that occasionally exceeds federal standards for eight heavy metals when tested using the standard EPA toxicity evaluation procedure. The current protocol for making this determination is the Toxicity Characteristic Leaching Procedure, or TCLP, EPA SW 846-1311. The TCLP and other leaching procedures evaluate the solubility of any heavy metals contained in a material, as opposed to the total concentrations of elements that are present in the material.

A heavy metal that is in a higher state of solubility, such as a metallic oxide or hydroxide, is more likely to dissolve and possibly impact surface or ground water than a metal that is in a less soluble form, such as a metal silicate, should the pavement and/or aggregate break up and release fine particles. Chemical fixation, through the process of ion exchange, permanently converts the heavy metals in the ash to forms of low solubility metal silicates. After the low solubility metal silicates are formed, the use of pozzolans causes the encapsulation of the wastes onto particle surfaces, and metals are incorporated into a cementitious matrix.

During the demonstration phase of TAP product development, 100 tons of TAP were produced at the MSC/JTM pilot plant from ash supplied by a mass-burn facility in the Midwest.

Representative samples of the TAP were obtained daily over the 15 days of production. The samples were composited and then shipped to Minneapolis for leachability testing. Representative samples of the MSW ash had been obtained and composited at the waste-to-energy facility as the ash was being loaded into trucks for transport to Georgia for processing in the pilot facility.

The values shown in Table 2 are the averages of leachability testing of four representative composite samples of the "raw" combined ash used in the TAP production, and of the resulting TAP product.

TABLE 2 -- Results of leachability testing using TCLP procedure[4]

PARAMETER	RAW ASH (mg/l)	TAP (mg/l)	MCL[5] (mg/l)
Arsenic	<0.002	<0.05	0.05
Barium	2.93	0.55	1.00
Cadmium	1.07	0.0002	0.005
Chromium	0.024	0.063	0.10
Lead	5.6	<0.0002	0.015
Mercury	0.006	0.0003	0 002
Selenium	<0.005	<0.005	0.05
Silver	<0.01	<0.01	0.10
Copper	1.20	0.018	1.00
Zinc	120.25	0.01	5.00

The raw ash and TAP leachates were analyzed for the eight RCRA "priority" metals (Arsenic, Barium, Cadmium, Chromium, Lead, Mercury, Selenium, and Silver) plus Copper and Zinc, which are secondary contaminants. These results are compared to the Maximum Contaminant Levels (MCL's) contained in the Federal Drinking Water Standards, which are in the far right column of Table 2. The average values of 5.6 mg/l for lead and 1.07 mg/l for cadmium in the raw ash that was used in the TAP manufacture slightly exceeded the hazardous thresholds of 5.0 and 1.0 mg/l respectively.

To simulate the "worst-case" environmental scenario, i.e., complete disintegration of the aggregate, the TAP samples were crushed to a minus 100-mesh powder prior to the leaching procedure. The standard TCLP requires only that particles be reduced in size to under 3/8 inch (9.5 mm). Due to the alkalinity of the TAP, it was necessary to use Extraction Fluid No. 2, which is acetic acid with a pH of 2.88 ± 0.05. Extraction Fluid No. 2 is approximately 100 times more acidic than extraction Fluid No. 1, which has a

[4] Analyses performed by Braun Intertec Environmental, Minneapolis, MN.

[5] Maximum Contaminant Levels set by the U.S. EPA for primary and secondary Federal Drinking Water Standards for public water supplies.

pH of 4.88 ± 0.05, or acid rain, which has a pH of approximately 5.0.

It is apparent from a comparison of TAP results to raw ash results in Table 2 that there is a significant reduction in leachability of the metals analyzed due to processing, chemical fixation, and addition of binders.

Interestingly, there was a slight increase in chromium leachability from the raw ash to the TAP. Upon investigation, it was determined that the source of increased chromium levels was the portland cement that had been used as one of the binder materials. However, the chromium was not in the Chromium (VI) or hexavalent form, which is carcinogenic, but was in the Chromium (III), or trivalent form. Trivalent chromium is neither carcinogenic nor a significant threat to human health at these concentrations, and environmental standards are less rigid for this form.

DURABILITY OF TAP AGGREGATE

The TAP aggregate mix design was initially developed by subjecting trial mix pellets to a series of standard aggregate durability tests. Early in the project, the decision was made for competitive purposes to formulate the TAP mix design so that the synthetic aggregate would meet the same performance standards that are required for conventional natural aggregates. Because of MSC efforts that were ongoing in Minnesota at that time, the TAP was targeted to meet Minnesota Department of Transportation (MnDOT) physical requirements for aggregates to be used in bituminous wearing surfaces. The standardized MnDOT aggregate durability tests used were:

* The LA Abrasion Test, ASTM C-131;

* Soundness of Aggregates by Use of Magnesium Sulfate, ASTM C-88, and

* Soundness of Aggregates by Freezing and Thawing, AASHTO T-103, Procedure B.

Although gradation and percentages of impurities such as spall materials, lumps, and shale content are MnDOT aggregate requirements [3], it was assumed that proper gradation would be accomplished through crushing and addition of specific size fractions of natural aggregates. Spall materials, clayey lumps, and shale were not materials of concern in the manufactured TAP synthetic aggregate.

TABLE 3 -- Engineering properties of TAP aggregate[6]

Procedure	TAP	MnDOT
Los Angeles Abrasion (ASTM C-131)	25-35% loss	<40% loss
Soundness by Use of $MgSO_4$ (ASTM C-88)	1-3% loss	<10% loss[7] or <20% loss[8]
Freeze-Thaw (AASHTO T-103)	9-12% loss	<12% loss

The Los Angeles Abrasion and the Sulfate Soundness tests were run using the standard ASTM procedures. The MnDOT aggregate freeze-thaw test is the current Procedure B of AASHTO T-103. In Procedure B, the aggregate is first saturated with a 0.5% methyl alcohol-water solution in a vacuum chamber for 15 minutes. The aggregate is then placed one layer deep in a shallow pan containing 1/4 inch (6.4 mm) of the alcohol-water solution [4]. MnDOT requires 16 cycles of freezing and thawing, and considers a maximum loss of 12% to be acceptable.[9] Although the TAP is not as freeze-thaw resistant as many natural aggregates, losses have been reduced to the range of 9-12%, which is within generally acceptable limits.

Initial high losses on the freeze-thaw test pointed out the need to reduce the TAP absorption and/or increase the strength of the cementitious matrix. High absorption is a

[6] Analyses performed by Braun Engineering, Law Engineering, and GaDOT.

[7] Applies to Mn DOT Class D aggregate containing >45% non-igneous particles.

[8] Applies to Mn DOT Class D aggregate containing 20-45% non-igneous particles.

[9] Freeze-Thaw testing is not mandated by MnDOT, but is considered useful in evaluating new sources of aggregate supply. The 12% maximum loss is a "generally-accepted" indicator of acceptable freeze-thaw soundess by the Minnesota, Michigan, New York, and other Departments of Transporation.

characteristic of MSW ashes [5], and is manifested in other properties, especially higher asphalt demand for the TAP. The problem was addressed by coating the finished TAP product with a thin layer of lime, minimizing exterior pore space.

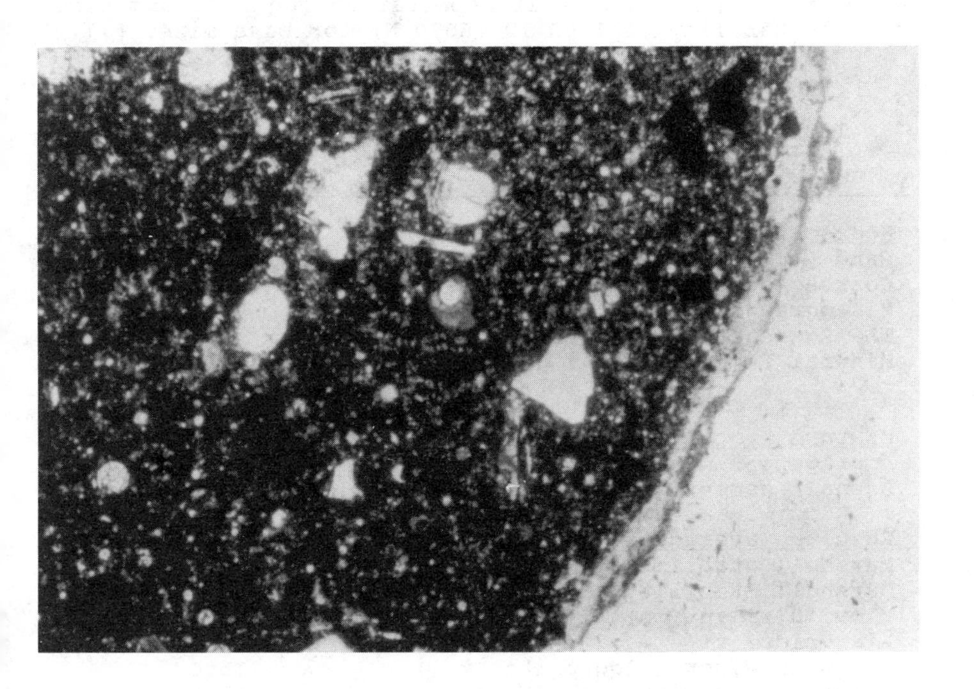

FIG. 2 -- TAP Sample at 10x magnification

Figure 2 is a photomicrograph of a cross-section of a TAP sample at 10x magnification that was prepared for petrographic examination. Note the coating on the surface of the particle. In addition to reducing asphalt demand, it is anticipated that the coating will also improve anti-stripping properties [6].

TAP AGGREGATE IN HOT-MIX ASPHALT

Preliminary experimentation was performed in the bituminous laboratory of Law Engineering in Atlanta, GA for the purpose of determining an approximate asphalt demand for a bituminous mixture utilizing the TAP aggregate.

During this preliminary bituminous mix design phase, the TAP aggregate was incorporated into the mix at a rate of 30% of the total aggregate mixture by weight. Marshall Stability results shown in Table 4 demonstrate that the TAP mixture tested above 2500 lb (11,120 N), a level comparable to test results for bituminous mixtures using conventional aggregates. The Georgia Department of Transportation requires a minimum Marshall Stability of 1500 lb (6670 N) for base mixes [7].

TABLE 4 -- Hot mix data using Marshall Stability Method before coating[10]

Aggregate Blend	
Sand	22.0%
Coarse Aggregate	21.0%
M1 Aggregate Screening	22.0%
TAP Aggregate	30.0%
Mineral Filler	5.0%
Total Aggregate	100.0%
Mixture Proportions	
Aggregate	92.5%
Asphalt cement[11], % by weight of total mixture	7.5%
	100.0%
Results (average of two samples)	
Marshall Stability - Measured, lb (N)	3109 (13,830)
Marshall Stability - Corrected, lb (N)	2597 (11,550)
Flow, 0.01 in (mm)	10 (2.5)
Air Voids, %	5.2
Voids in Mineral Aggregate, %	15.6
Voids Filled with Asphalt, %	66.3
Effective asphalt cement, %	5.4
Mixing temperature, °C	157
Bulk specific gravity	2.096

The high asphalt demand of 7.5% in this mix pointed out the need to reduce the absorption of the TAP aggregate. High asphalt contents affect both cost and performance of the material, and would hamper acceptance of TAP as a bituminous pavement aggregate. At this point, it was necessary to evaluate some different methods of reducing the TAP absorption.

---

[10] Analysis performed by Law Engineering, Atlanta, GA.

[11] AC-30, viscosity graded asphalt cement.

The method selected was to coat the TAP particles with lime. This had the effect of sealing many of the surface pores on the TAP, while also acting as an anti-stripping agent.

Table 5 shows the gradations of the aggregate materials used in the later mixes. The TAP and natural aggregate were combined to meet the Georgia Department of Transportation aggregate gradation requirements for a "Type E" bituminous mix [6]. In this mixture, the material coarser than a No. 4 sieve is TAP, while the material finer than a No. 4 sieve is natural aggregate. The natural aggregate was crushed granitic gneiss from the Vulcan Materials, Norcross, GA quarry.

TABLE 5 -- Aggregate size gradations: As-received and as used in mixtures[12]

SIEVE SIZE	TAP % PASSING	NATURAL AGGREGATE % PASSING	MIXTURE % PASSING	GaDOT TYPE E REQUIRED % PASSING
3/4"	100	100	100	100
1/2"	94	100	93	85-100
3/8"	79	100	78	70-85
No. 4	13	79	62	--
No. 8	2	62	46	44-48
No. 16	1	50	36	--
No. 30	1	40	27	--
No. 50	1	27	18	13-22
No. 100	1	14	12	--
No. 200	0.5	7	6	4-7

Table 6 provides results of the second series of tests, using four asphalt mixtures containing the TAP aggregate coated with lime. A cross-section of an asphalt concrete specimen made during this series of tests is shown in Figure 3. Four different asphalt cement contents were tested. Mixes 2 and 3, with asphalt cement contents in the range of 5.7 to 6.2 percent by weight of total mixture, met the Georgia DOT "Type E" requirements for stability, air void

12 Analyses performed by Law Engineering, Atlanta, GA.

content, and percent aggregate voids filled. The flow value of 7 in Mixes 2 and 3 was slightly low, as GaDOT requires a flow of 8-16 (2.03-4.06 mm). Low flow values can be addressed through a modification in aggregate gradation or a small increase in alsphalt cement content.

TABLE 6 -- Asphaltic concrete mixture properties after coating

	Mix # 1	Mix # 2	Mix # 3	Mix # 4
Asphalt Cement Content, % of Total Mix	5.2	5.7	6.1	6.6
Bulk Specific Gravity Dry	2.169	2.183	2.187	2.192
Theoretical Maximum Specific Gravity (Gmm)	2.311	2.295	2.282	2.269
Air-Voids, %	6.2	4.9	4.2	3.4
Voids Filled w/Asphalt, %	64.1	71.1	75.6	80.4
Marshall Stability, lb (N)	2150 (9560)	2540 (11,300)	2500 (11,120)	2140 (9520)
Flow, 0.01 in (mm)	4 (1.02)	7 (1.78)	7 (1.78)	10 (2.54)

The results presented in Table 6 are averages of three laboratory compacted specimens per asphalt cement content, except stability and flow for which an average of two specimens per asphalt cement content was used. Specimens were compacted using 50 blows per side. Theoretical maximum specific gravity (Gmm) was measured in accordance with ASTM D 2041 for the 5.7% asphalt cement mixture. Gmm's for the other mixtures were calculated based on Gmm for 5.7% asphalt cement and the asphalt having a specific gravity of 1.037.

Traditionally, state highway agencies prefer that aggregate for use in asphaltic concrete be angular with flat surfaces in order to achieve good Marshall Stability values and resistance to rutting. The results of this series of tests indicate that stability values well above minimum can be achieved with rounded particles, although it is anticipated that a certain amount of flat surfaces will be produced if crushing of the TAP aggregate is performed to achieve gradations.

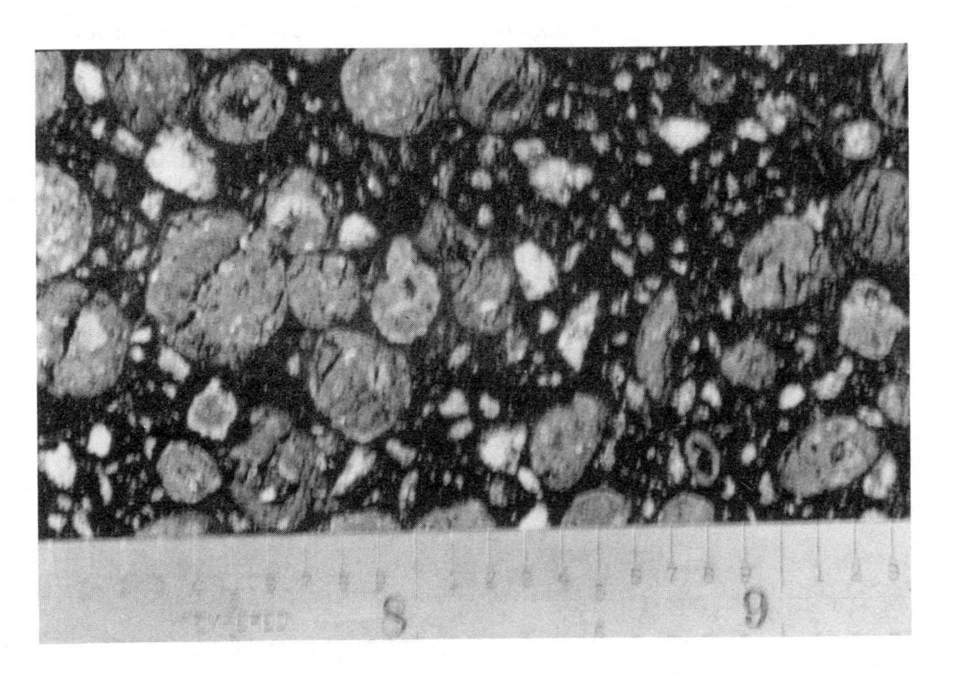

FIG. 3 - Cross-section of an asphaltic concrete specimen made with the TAP aggregate. This mix design achieved a Marshall Stability of 2540 lb (11,300 N) as shown in Table 6.

CONCLUSIONS

Results of the various laboratory tests performed to date indicate that the TAP aggregate should perform well in bituminous pavement. As the durability tests for qualification of aggregates in asphalt concrete are similar to those employed to qualify aggregates for use in portland cement concrete, these results demonstrate potential for the use of TAP in both applications.

Good Marshall Stability values averaging 2540 lb (11,300 N) were attained using an asphalt cement content of 5.7%, calculated as a percent of the total mix, after a modification was made to the aggregate manufacturing process to reduce absorption.

At this point, a field demonstration is the next step, to verify durability of the material under actual traffic loads and cyclic weather conditions.

To conclude, municipal solid waste combustor ash shows good potential for processing into useful materials of construction. It has been demonstrated that a carefully designed TAP product can meet two types of standards: engineering requirements as a construction material and stringent environmental standards.

REFERENCES

[1] U.S. Environmental Protection Agency, Characterization of Municipal Solid Waste in the United States: 1992 Update, National Technical Information Service, Springfield, VA, 1992.

[2] Integrated Waste Services Association, 1992.

[3] Minnesota Department of Transportation, Supplemental Specifications to the 1988 Standard Specifications for Construction, Specification 3139, "Graded Aggregate for Bituminous Mixtures", St. Paul, Minnesota, 1991.

[4] American Association of State Highway and Transportation Officials, Standard Specifications for Transportation Materials and Methods of Sampling and Testing, "Standard Method of Test for Soundness of Aggregates by Freezing and Thawing", AASHTO Designation: T 103-83 (1990), 15th Edition, 1990.

[5] Gress, David, et al, "The Use of Bottom Ash in Bituminous Mixtures", in Proceedings of the Fourth International Conference on Municipal Solid Waste Combustor Ash Utilization, edited by W.H. Chesner and F.J. Roethel, Arlington VA, 1991.

[6] Georgia Department of Transportation, Standard Specification - Construction of Roads and Bridges, Section 828 - "Hot Mix Asphaltic Concrete Mixtures", Atlanta, GA, 1983 Edition.

Roofing Shingles

David E. Newcomb,[1] Mary Stroup-Gardiner,[1] Brian M. Weikle,[1] and Andrew Drescher[1]

PROPERTIES OF DENSE-GRADED AND STONE-MASTIC ASPHALT MIXTURES CONTAINING ROOFING SHINGLES

REFERENCE: Newcomb, D.E., Stroup-Gardiner, M., Weikle, B.M., and Drescher, A., **"Properties of Dense-Graded and Stone-Mastic Asphalt Mixtures Containing Roofing Shingles,"** Use of Waste Materials in Hot-Mix Asphalt, ASTM STP 1193, H. Fred Waller, Ed., American Society for Testing and Materials, Philadelphia, 1993.

ABSTRACT: Fiberglass-backed and felt-backed roofing shingle wastes from manufacturing were introduced into dense-graded and stone-mastic asphalt mixtures. The common materials used in roofing and hot-mix applications such as asphalt, fine aggregate, and mineral filler suggest compatibility. It is hypothesized that possible improvements in the tensile characteristics of asphalt pavement mixtures might result from the fibers used in the roofing materials. Initially, three levels of roofing waste percentages were evaluated in each mixture type along with two grades of asphalt cement. The fiberglass shingles proved to be more susceptible to changes in asphalt content than the felt. As the percentage of roofing waste increased, the compactibility of the mixtures was increased. Resilient modulus and splitting tensile strength data further illustrate the differences between the modified mixtures and the control materials.

KEYWORDS: fiber reinforced hot-mix asphalt, roofing shingle waste, stone-mastic asphalt, split-mastic asphalt

INTRODUCTION

Background

It is estimated that roofing shingle production generates approximately 432,000 tons of waste annually in the United States, and about 36,000 tons of this is in the Twin Cities Metro Area of Minnesota. Another 8.5 million tons of waste material come from the rebuilding of shingle or hot-mop roofs. Disposal of this waste material is usually accomplished by transporting and depositing it in landfills. If a suitable means of reusing these materials can be found, then their environmental liability could be significantly reduced.

Since asphalt roofing shingles are comprised of approximately 35 percent asphalt, 45 percent sand, and 20 percent mineral filler, an alternative to landfill deposition is to use the roofing waste in a related bituminous material. Such applications could include its use in granular base stabilization, patching materials, or in hot-mix asphalt concrete [1,2]. In this paper, the use of roofing wastes in dense-graded

[1]Assistant professor, research fellow, research assistant, and professor, respectively, Civil & Mineral Engineering, University of Minnesota, 500 Pillsbury Dr. SE, Minneapolis, MN 55455.

and gap-graded asphalt mixtures will be examined with respect to their effects on mixture behavior and properties.

Researchers at the University of Nevada-Reno investigated the economic and technical aspects of using waste roofing from reconstruction in hot-mix asphalt [3,4]. They concluded that the use of shingle waste should result in a lower cost paving material. Paulsen, et al. [4] stated that the use of roofing waste tended to increase the stiffness of mixtures. This could be reasonably expected due to the use of higher viscosity asphalt in the shingles along with the reinforcing effect of the fiber. They suggested that the properties of the shingle asphalt and the gradation of the shingle aggregates be considered when formulating hot-mix asphalt. They further stated that up to 20 percent shingle material by volume (10 to 12 percent by weight) could be accommodated in the mixtures without detrimental effects.

While the Reno study focused on the use of construction roofing waste in dense-graded mixtures, the idea of using manufacturing waste in dense- or gap-graded asphalt mixtures is a relatively new notion. Stone mastic asphalt is a concept which has recently gained widespread publicity in the United States [5]. Originally developed in Germany in the 1960's as a means of combating studded tire wear, the idea was widely adopted across Europe for rut-resistant overlays [6]. The idea is to create stone-to-stone contact in the coarse aggregate, and bind it together using a mastic of a relatively hard asphalt cement, fine aggregate, and a modifier to prevent the asphalt from draining. The modifier may be either a polymer or a fiber [7]. Since roofing shingles contain a stiff binder as well as fibers, it was hypothesized that they could be used in place of the conventional modifiers.

Objective

The objective of this study was to evaluate the use of roofing shingle waste from the manufacturing process in hot-mix asphalt concrete mixtures. In dense-graded mixtures, it was hypothesized that the waste material might serve as a binder extender as well as a fiber reinforcement. In the stone mastic asphalt (SMA), it could serve as the binder stiffener to prevent drain down by replacing the fibers or mineral fillers commonly used in these mixtures.

Scope

The treatment of dense-graded and SMA mixtures can viewed as two experiments, because of the different considerations in formulating each of them. The dense graded mixture evaluation included two grades of asphalt cement, one gradation, three levels of roofing shingle content (0, 5.0, and 7.5 percent by weight of aggregate), and two roofing waste types (fiberglass and felt-backed). In the stone mastic asphalt mixtures, one asphalt cement grade, one aggregate gradation, one roofing shingle content, and three types of fiber additives were used. The control material for the SMA mixtures contained a commercial cellulose fiber.

MATERIALS

Asphalt

The neat asphalt cements added to the mixtures were 85/100 and 120/150 penetration grade [Standard Specification for Penetration-Graded Asphalt Cement for Use in Pavement Construction, ASTM D 946] materials obtained from the Koch Refinery in Inver Grove Heights, Minnesota. Only the 85/100 penetration grade asphalt cement was used in the SMA mixtures, since the literature indicated the need for stiffer binders in these. Table 1 shows the viscosities of the materials at 60°C and 135°C, before and after aging in a rolling thin film oven. The 85/100 grade asphalt has a viscosity which is slightly lower than is required for an AC-20 grade [Standard Specification for Viscosity-Graded Asphalt Cement for Use in Pavement Construction, ASTM D 3381], and the 120/150 penetration grade could be classified as an AC-10 viscosity grade [ASTM D 3381].

TABLE 1--Neat asphalt cement properties.

Penetration Grade	85/100	120/150
Viscosity, original		
60°C, P	1588	908
135°C, cSt	362	259
Viscosity, after RTFO		
60°C, P	5372	2804
135°C, cSt	275	449

Aggregates

The different gradations used for the dense-graded and SMA mixtures are shown in Figure 1. The dense gradation falls approximately in the middle of the specification band for a Minnesota Department of Transportation type 2341 mixture [8]. The SMA gradation is suggested by the German Federal Department of Transportation [6]. In both cases, the maximum aggregate size is 12.5 mm.

The dense gradation is comprised of aggregates from two sources. The major portion (76 percent by weight) of the blend is a partially crushed river gravel from the Commercial Asphalt, Inc. pit located in Lakeland, Minnesota. The portion of the blend larger than 9.5 mm in size was a granite obtained from Meridian Aggregates in Granite Falls, Minnesota. In order to prevent aggregate gradation from becoming a covariable in the experiment, the dense-graded material's composition was adjusted for the mineral material content in the roofing shingles. The gradation of the mineral filler and ceramic coated aggregate was supplied by Certainteed Corporation in Shakopee, Minnesota.

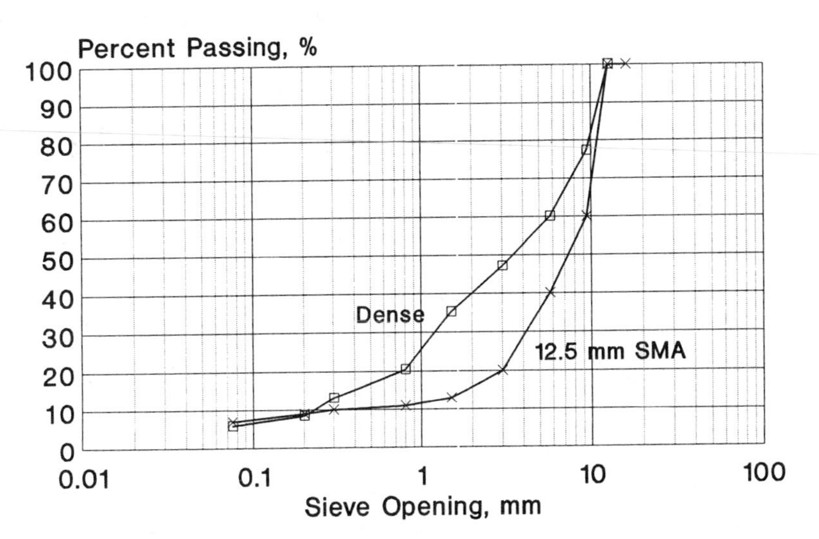

FIG. 1--Aggregate gradations.

TABLE 2--Aggregate properties.

Mixture Type	Dense-Graded	SMA
Bulk Specific Gravity	2.63	2.51
Bulk Specific Gravity, SSD	2.67	2.55
Apparent Specific Gravity	2.75	2.62
Absorption Capacity, %	1.67	1.70

The SMA gradation was also a blend of the river gravel and crushed granite aggregates. However, in this case, the river gravel use was restricted to the fine aggregate portion of the blend. All material greater than 4.75 mm was comprised of the granite. The properties of the SMA aggregate are shown in Table 2.

Roofing Waste

The roofing waste was generated by the Certainteed Corporation's Shakopee, Minnesota facility. The waste was transported to Omann Brothers in Rogers, Minnesota for processing. The waste was ground by two hammermills in tandem, water cooled, and stockpiled. Water-cooling after grinding was considered necessary to prevent the material from agglomerating. It also created high stockpile moisture contents; 3.8 and 10.3 percent for fiberglass and felt, respectively. In the laboratory work, the material was dried under a fan at ambient temperature over a 12-hour period. However, potential compaction and moisture sensitivity problems in field mixtures could be created by this condition.

The ground roofing waste had a size range of about 5 to 30 mm, although agglomeration of the particles made it impossible to perform a gradation analysis on the material. The specific gravity of the roofing waste material was determined using a modification to ASTM procedure C128, Standard Test Method for Specific Gravity and Absorption of Fine Aggregate. The modification involved applying a partial vacuum to the material in order to remove entrapped air. The specific gravities were about 1.29 for the felt-backed material, and 1.37 for the fiberglass shingles.

Cellulose Fiber

The cellulose fiber used in the SMA control mixture is marketed under the tradename Arbocel and is produced by J. Rettenmaier and Sohne of Germany. The material has a cellulose content of between 75 and 80 percent, and a bulk density of 25 to 30 g/l. The average fiber length is 1100 μm, and the average diameter is 45 μm [9].

MIXTURE DESIGN

The optimum added asphalt content for all mixtures was determined using the Marshall method of mix design [Standard Test Method for Resistance to Plastic Flow of Bituminous Mixtures Using Marshall Apparatus, ASTM D1559]. The shingle materials were at ambient when they were added to the mixtures. They were introduced into the mixtures during the mixing process after the aggregate had been initially coated. The roofing waste showed no problems in dispersing into the mixtures; noticeable pockets of roofing waste were not present in the final mixture. After mixing, the loose material was placed in a 135°C oven for three to four hours for short-term aging. Compaction was achieved using a rotating-base, bevel-head Marshall hammer, applying 75-blows per side for dense-graded mixtures and 50 blows per side for the SMA mixtures.

In order to understand the effect of the roofing waste on the total binder content, extractions were performed on selected dense-graded mixtures containing the felt material after mixing. These results are shown in Table 3. At the 2.5 percent roofing waste level, the mixture gained about 1.5 percent asphalt, and the gain was 2.7 percent when the roofing waste content was increased to 7.5 percent. The increase in asphalt contents for these mixtures is consistent with previously reported experience in fiber reinforced asphalt concrete mixtures [10].

TABLE 3--Effect of felt-backed roofing waste on binder content for dense-graded mixtures.

Optimum Added Asphalt Content, %	4.3	3.6	3.7
Roofing Waste Content, %	2.5	5.0	7.5
Total Binder Content, %	4.8	5.0	6.4

In order to mimic field density of the dense-graded mixtures, it was necessary to define the compactive effort which would result in an air void content of between six and eight percent at the optimum asphalt content. Figures 2 and 3 show how the air void contents varied with Marshall compaction for the dense-graded mixtures. It can be seen that for the felt-backed roofing shingles, the mixture with higher concentration of shingles tends to compact more readily (Figure 2), while the difference for the fiberglass shingles is not as great (Figure 3).

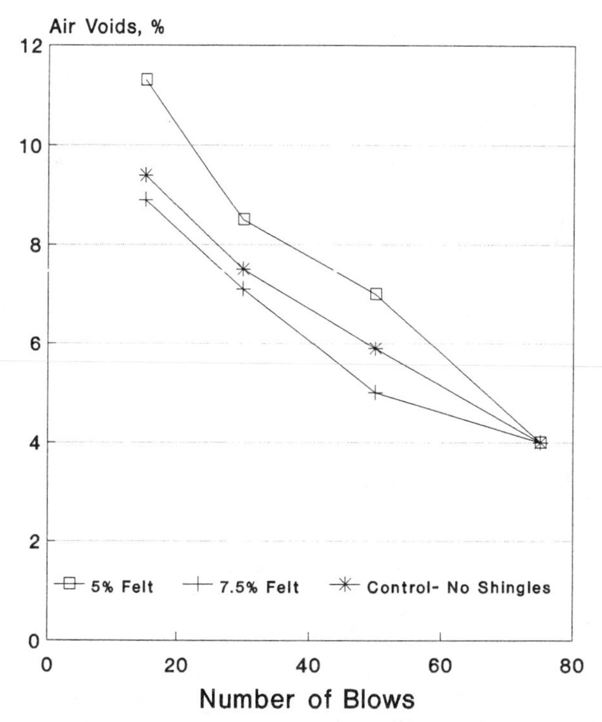

FIG. 2--Air voids versus Marshall compactive effort for felt-backed material in dense-graded mixture.

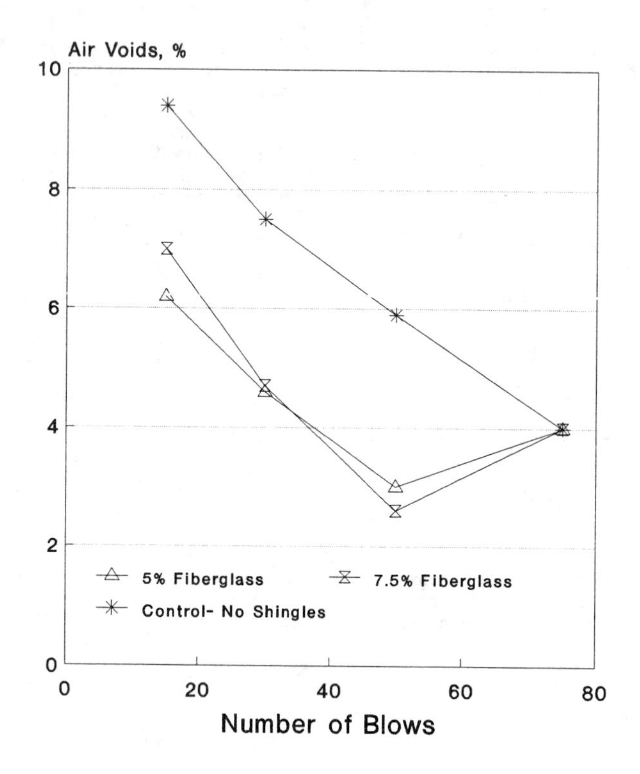

FIG. 3--Air voids versus Marshall compactive effort for fiberglass-backed material in dense-graded mixture.

The SMA mixtures were subjected to a compactive effort of 50 blows per side. The waste shingle content used in the SMA mixtures was fixed at 10 percent by weight of the aggregate. The control material for this type of mixture contained cellulose fibers to stiffen the binder. As per the manufacturer's recommendation, the fibers were added at a rate of 0.3 percent by weight of mix.

MIXTURE EVALUATION

Testing

Resilient modulus and indirect tensile testing were done in accordance with ASTM D4123, Standard Test Method for Indirect Tension Test for Resilient Modulus of Bituminous Mixtures, using a closed-loop hydraulic test system with a 10-kN capacity. The resilient modulus test was accomplished using a 1 Hz frequency with a 0.1-s load application followed by a 0.9-s rest period. The resilient modulus was calculated using the total recoverable strain, and Poisson's ratios of 0.2, 0.35, and 0.5 for temperatures of 1, 25, and 40°C, respectively. Values of Poisson's ratios were selected based upon the Strategic Highway Research Program recommendations for testing Long-Term Pavement Performance materials.

Indirect tensile strength tests were performed at -18°C and 25°C using vertical displacement rates of 0.012 and 50 mm/min, respectively. The lower displacement rate was chosen for the colder temperature in order to ascertain the effects of creep tensile loading corresponding to lower rates of loading in a pavement subjected to a temperature change.

Results

The results of the resilient modulus testing and indirect tensile strengths for all mixtures are listed in Table 4. The values shown are generally the average of three tests. The coefficient of variation within a set of three specimens was not more than eight percent for resilient modulus, and it was not more than 13 percent for indirect tensile strength.

Dense-Graded Mixtures--The resilient modulus versus temperature curves for dense-graded mixtures made with 85/100 penetration asphalt are shown in Figure 4. The most notable feature of this graph is that the control mixture resilient modulus is 1.5 to two times greater at 1°C than those containing the roofing wastes. At 25°C, the control mixture has a resilient modulus which is consistent with the shingle modified mixtures, and at 40°C, it is slightly stiffer than all except the mixture containing 5 percent felt roofing waste. In general, it can be seen that mixtures containing 5 percent shingles are stiffer than those containing 7.5 percent at 25 and 40°C, and that mixtures containing felt roofing waste are stiffer than those with fiberglass-backed shingles. The softer behavior of the modified mixtures is most likely due to the increased binder content caused by the asphalt in the roofing waste, although it appears as though the temperature susceptibility is decreased by the inclusion of the waste material. Similar behavior was noted for mixtures containing the 120/150 penetration grade asphalt cement.

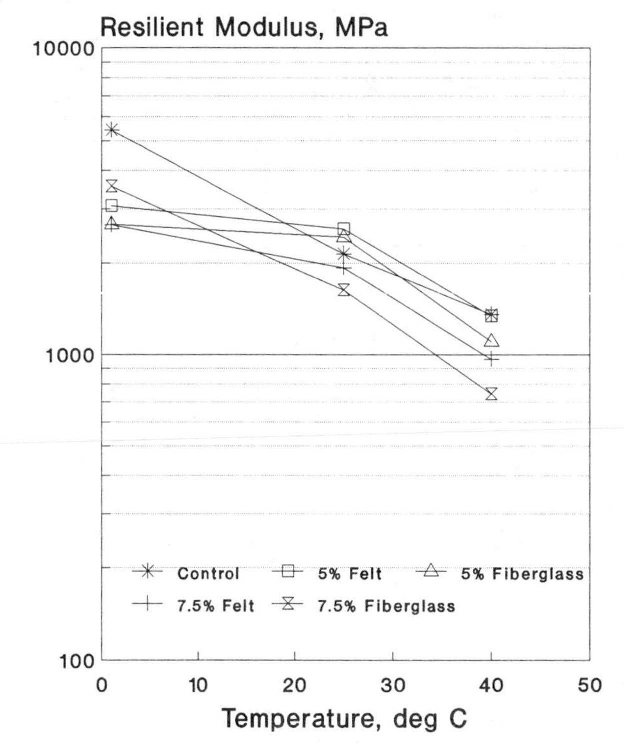

FIG.4--Resilient modulus versus temperature for dense-graded mixtures with 85/100 penetration grade asphalt.

TABLE 4--Resilient modulus and tensile strength results.

Asphalt Cement	Shingle Type	Percent Shingles	Resilient Modulus, MPa			Tensile Strength, kPa	
			1°C	25°C	40°C	-18°C	25°C
120/150	Felt	0	5133	2376	1223	2659	643
		5	2905	2260	1222	2312	787
		7.5	2411	1226	591	1549	486
	Fiberglass	0	5133	2376	1223	2659	643
		5	4062	2028	849	1972	432
		7.5	4070	1599	871	1882	431
85/100	Felt	0	5420	2140	1355	3234	908
		5	3076	2584	1341	1875	890
		7.5	2669	1930	968	1420	587
	Fiberglass	0	5420	2140	1355	3234	908
		5	2680	2429	1107	2482	465
		7.5	3556	1630	744	1855	387
SMA	Control	0	7063	2501	935	2746	927
	Felt	10	6927	2259	894	2555	840
	Fiberglass	10	6681	2579	1300	3275	826

Note: Values are the average of three tests.

The effect of roofing shingle concentration and type on the 25°C indirect tensile strength of dense-graded mixtures is shown in Figure 5. One should first note the tensile strengths of the control mixtures (0 percent shingles) for the 85/100 and 120/150 penetration asphalts. The control mixture for the 85/100 has a tensile strength of almost 1.5 times that of the 120/150 control mixture. Adding the shingles to the 85/100 asphalt tended to decrease the 25°C tensile strength, while the tensile strength increased significantly for the 120/150 asphalt mixtures up to the level of 5 percent shingle concentration for the felt material. It would seem that the additional asphalt contained in the felt shingle material contributed to the slight weakening of the 85/100 mixture, while possibly the stiffer roofing asphalt and the felt fibers in the shingle material strengthened the 120/150 asphalt mixtures. There does not appear to be a great deal of difference in tensile strengths between the asphalt cement types as between roofing shingle types at the five and 7.5 percent levels. In other words, the fiberglass mixtures all had tensile strengths between 387 and 465 kPa. The mixtures made with the felt-backed shingles seemed a little more sensitive to the grade of added asphalt, but the concentration of the shingle material seems to control the tensile strength of the mixtures somewhat.

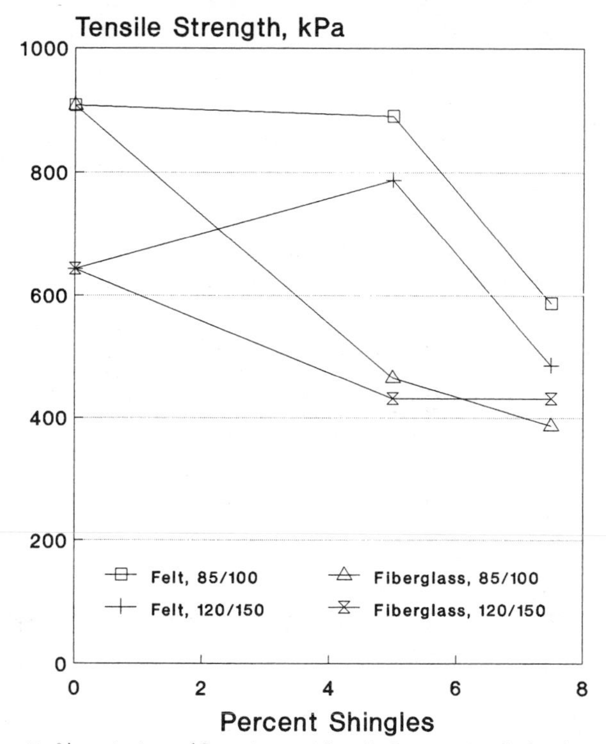

FIG.5--Indirect tensile strength of dense-graded mixtures at 25°C.

Figure 6 shows how the increase in shingle material content decreases the tensile strength of the mixtures at low temperature. Here it can be seen that mixtures made with the felt-backed roofing material generally have lower tensile strengths for the same asphalt grade than those prepared with fiberglass roofing waste. The decrease in tensile stress at -18°C is slightly less dramatic for the samples prepared with the 120/150 penetration asphalt.

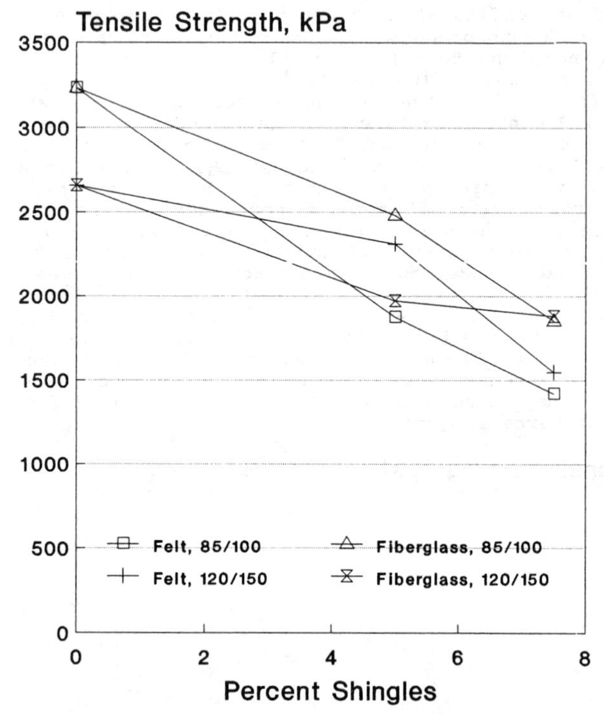

FIG. 6--Indirect tensile strength of dense-graded mixtures at -18°C.

Typical tensile stress versus tensile strain curves at -18°C for the 120/150 control mixture and the 7.5 percent fiberglass mixture are shown in Figure 7. This illustrates the embrittlement of the mixture at low temperatures when fiberglass roofing material is used in the mixture. The fiberglass waste produced a mixture which had a significantly lower peak stress as well as an overall lower strain capability.

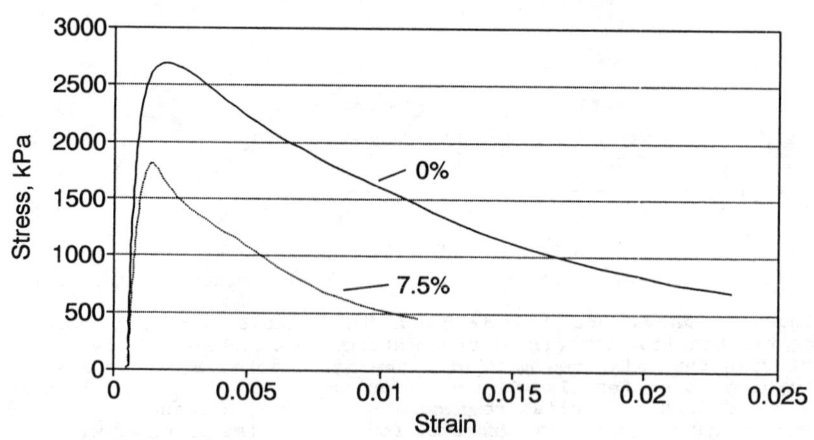

FIG. 7--Tensile stress versus tensile strain for 120/150 dense-graded mixtures.

Stone Mastic Asphalt--Figure 8 shows the resilient modulus test results for the SMA mixtures. It can be seen that all the mixtures behave consistently over the range of temperatures from 1 to 40°C. At 1°C, the SMA materials all have a resilient modulus of about 7,000 MPa, and at 25°C, the mean is approximately 2,500 MPa. The fiberglass-backed roofing SMA shows a slightly stiffer behavior at 40°C than either the control or felt-backed shingle SMA.

Indirect tensile strengths at 25°C for the SMA mixtures are shown in Figure 9. It can be seen that the shingle-modified materials had tensile strengths which were approximately 100 kPa lower than the SMA mixture prepared with cellulose fibers. The lower strength may be attributable to the extra asphalt contributed by the shingle material. This may not be completely detrimental as the added asphalt may contribute to a greater ductility in the shingle waste SMA's.

Figure 10 shows that the mixture containing fiberglass shingle waste had the highest tensile strength at -18°C, whereas the control and felt-backed mixtures were relatively close in terms of their low temperature strengths. However, an examination of typical stress-strain curves for these materials at low temperature (Figure 11) shows that the felt shingle material is more ductile than either control mixture or the fiberglass mixture. This behavior could be due to a combination of the added asphalt from the shingle material and the interlocking of the felt fibers. The added ductility seen in the felt material may be beneficial in providing resistance to low temperature cracking.

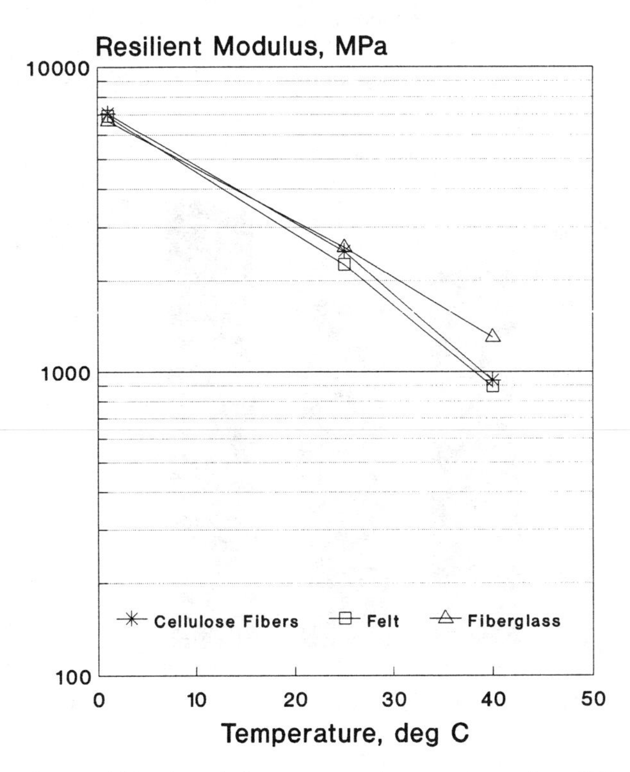

FIG.8--Resilient modulus versus temperature for SMA mixtures.

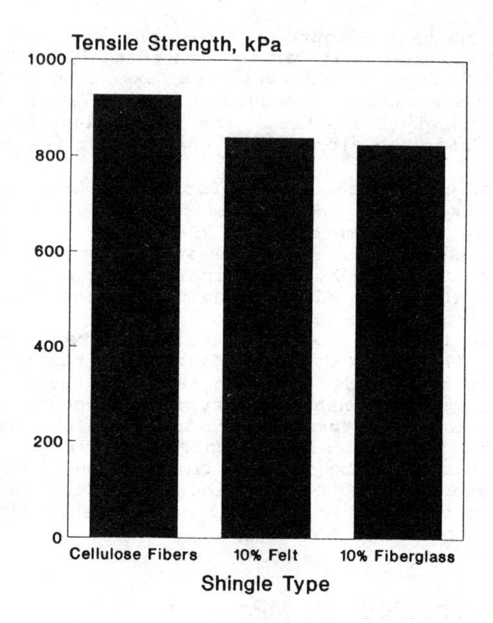

FIG.9--Indirect tensile strengths at 25°C for SMA mixtures.

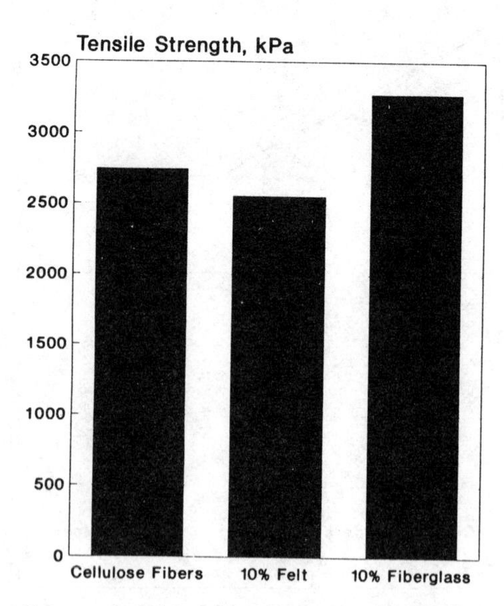

FIG. 10--Indirect tensile strengths at -18°C for SMA mixtures.

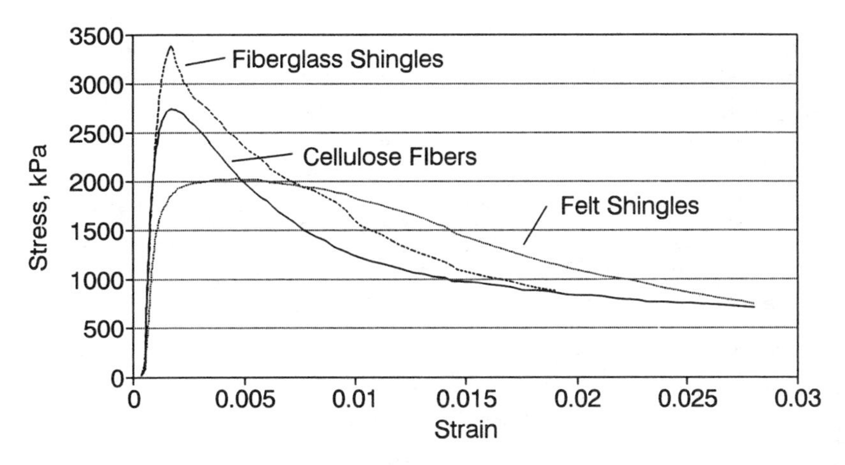

FIG. 11--Tensile stress versus tensile strain at -18°C for SMA mixtures.

CONCLUSIONS

Based on the data presented in this paper, the following conclusions are made:

Dense-Graded Mixtures

1. Manufactured roofing shingle waste can be incorporated into dense-graded asphalt concrete.
2. The use of roofing waste in such mixtures can lower the resilient modulus at low temperatures as well as at high temperatures. Overall, the roofing waste mixtures exhibited less temperature susceptibility.
3. The tensile strength of dense-graded mixtures at 25°C was maintained or increased with the addition of 5 percent felt shingles, and was decreased when this amount was increased to 7.5 percent. Fiberglass shingles tended to decrease tensile strength at either the 5 or 7.5 percent concentration level.
4. At -18°C, the tensile strength of the roofing waste modified mixtures decreased as the percentage of roofing waste increased.

SMA Mixtures

1. Up to 10 percent manufactured roofing waste can be used in a stone-mastic application.
2. The resilient modulus of the three SMA mixtures did not vary significantly at 1 or 25°C. However, the fiberglass shingle material had a greater resilient modulus at 40°C.
3. Tensile strengths at 25°C for the SMA mixtures containing shingle wastes were reduced about 10 percent compared to the mixture containing cellulose fibers.
4. At low temperature (-18°C), the fiberglass shingle SMA had the highest peak tensile stress and the most brittle behavior. The felt shingle SMA had the lowest peak tensile stress and the most ductile behavior of the three SMA mixtures.

SUMMARY

In this paper, it has been shown that the use of manufactured roofing waste affects the tensile behavior of asphalt concrete mixtures. This is due to the added asphalt contributed by the shingle material as well as the presence of fibers in the shingles. Further research is being conducted regarding the permanent deformation characteristics of such mixtures. Additionally, waste shingles from building reconstruction projects are being evaluated. It is clear that the type and nature of roofing waste must be accounted for when using these materials in asphalt paving mixtures, and that a laboratory evaluation of the tensile properties of the mixtures should be a part of the process.

ACKNOWLEDGEMENTS

This work is being performed under contract to the Minnesota Department of Transportation with funding from the Minnesota Office of Waste Management. The authors gratefully acknowledge the support and technical advice provided by Roger Olson of Mn/DOT.

REFERENCES

[1] Klemens, T.L., "Processing Waste Roofing for Asphalt Cold-Patches," Highway and Heavy Construction, Vol. 134, No. 5, April 1991, pp. 30-31.

[2] "Technical Data Sheet: ReACTS - HMA," ReCLAIM, Inc., Tampa, FL, 1991.

[3] Epps, J.A. and Paulsen, G., Use of Roofing Wastes in Asphalt Paving Mixtures: Economic Consideration, Report No.6-331-709-2, Center for Construction Materials, Civil Engineering Department, University of Nevada-Reno, 1986.

[4] Paulsen, G., et al., Roofing Waste in Asphalt Paving Mixtures, Report No. 6-331-709-1, Center for Construction Materials, Civil Engineering Department, University of Nevada-Reno, 1986.

[5] Warren, J.M., "SMA Comes to the USA," Hot Mix Asphalt Technology, Vol.6, No. 2, National Asphalt Pavement Association, Lanham, Maryland, 1991, pp. 5-9.

[6] Bellin, P.A.F., "Use of Stone Mastic Asphalt in Germany: State of the Art," preprint, Transportation Research Board, Washington, DC, 1991.

[7] Decoene, Y., "Contribution of Cellulose Fibers to the Performance of Porous Asphalts," Transportation Research Record No. 1265, 1991, pp. 82-86.

[8] Minnesota Department of Transportation, Standard Specifications for Construction, 1983 edition, 1983, pp. 249-250.

[9] "Arbocel Asphalt," product pamphlet, J. Rettenmaier & Sohne, D-7092 Ellwangen-Holzmuhle, Germany.

[10] Freeman, R.B., et al., "Polyester Fibers in Asphalt Paving Mixtures," Asphalt Paving Technology 1989, Vol. 58, Association of Asphalt Paving Technologists, 1989, pp. 387-409.

Kenneth F. Grzybowski[1]

RECYCLED ASPHALT ROOFING MATERIALS--A MULTI-FUNCTIONAL, LOW COST HOT-MIX ASPHALT PAVEMENT ADDITIVE

REFERENCE: Grzybowski, K. F., **"Recycled Asphalt Roofing Materials--A Multi-Functional, Low Cost Hot-Mix Asphalt Pavement Additive,"** Use of Waste Materials in Hot-Mix Asphalt, ASTM STP 1193, H. Fred Waller, Ed., American Society for Testing and Materials, Philadelphia, 1993.

ABSTRACT: A low cost (less than domestic cellulose fibers), commercially available recycled asphalt roofing material, ReACT's HMA™ [2], with a uniform gradation (minus No. 20 sieve, plus No. 70 sieve), was evaluated as an additive in conventional dense-graded hot mix asphalt designs to assess the overall efficacy. Additionally, these same materials were pre-blended with neat asphalt cement binders to assess their applicability in Stone Mastic Asphalt mix designs. It was hypothesized the components of asphalt roofing materials, if properly processed/recycled, would interact synergistically with the asphalt pavement binder and aggregate to yield results similar to current in-use modifiers and additives, such as polymers, cellulose fibers, mineral fillers, and low-penetration natural asphalts. Field trials indicate the recycled asphalt roofing materials can be processed into conventional hot-mix asphalt designs, much like recycled asphalt pavement (RAP). Results of experimental mix designs indicated neat asphalt savings of up to 50% are achievable. Experimental mixes exhibited improved high-temperature susceptibility and rut resistant properties, while allowing for total binder contents up to 7-8 weight percent. Evaluation of recycled asphalt roofing material pre-blended with neat asphalt cement indicates these materials are a functional equivalent to currently used binder modifiers, and their successful use in stone

---

[1] President, PRI Asphalt Technologies, Inc., 5310 56th Commerce Park Boulevard, Tampa, FL 33610.

[2] ReACT's HMA™, a registered trademark of ReClaim, Inc., Tampa, Florida.

mastic asphalts is highly probable. Overall, the recycled asphalt roofing materials are multi-functional in pavement mixtures, and their use can both alleviate environmental landfill concerns and result in cost-effective improved pavement designs.

KEYWORDS: asphalt roofing waste, stone mastic asphalt, Marshall Properties, SHRP Bending Beam Rheometer, drain-down, asphalt modifiers, rut resistance, recycling, environmental

---

BACKGROUND

It is estimated 11,000,000 tons of asphalt roofing waste is generated annually in the United States. The greatest single source, estimated at 10,000,000 tons annually, is from roof replacements. [1] A second large source is manufacturers of prepared roofing materials, and is estimated at 1,000,000 tons per year. It is further estimated that without recycling alternatives, the roofing wastes create 22,000,000 cubic yards of an environmental landfill burden.

The composition of the roofing waste materials is complex, varying between product types and forms, but has been defined through sample evaluation programs, as 30-40 weight percent air-blown asphalt, 50-60 weight percent inorganic mineral fillers/granules and 1-12 weight percent inorganic and/or organic fiber.

Asphalt roofing shingles comprise the dominant tonnage produced annually, 10.8 million tons, versus all prepared roofing products manufactured, 11.8 million tons. [1] Figure 1 displays a basic asphalt roofing shingle component profile. These shingles are manufactured in sheet form, 30.5 cm wide x 91.4 cm long x 32 mm thick, and have typical component properties as displayed in Table 1.

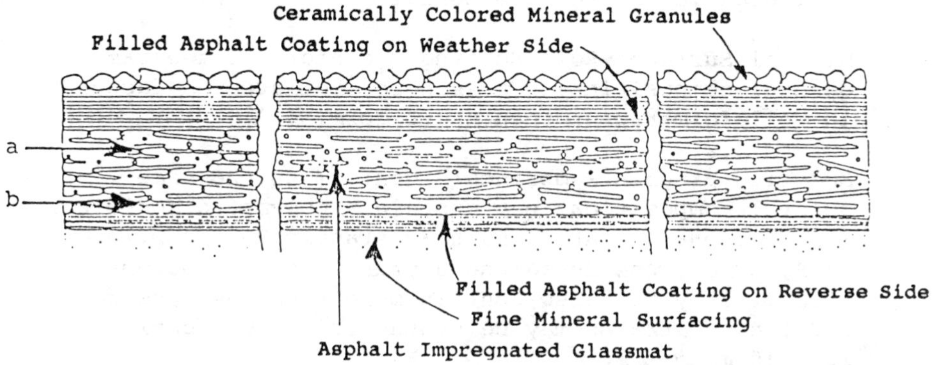

FIG. 1--Profile of a residential roofing shingle.

TABLE 1--Components & properties of asphalt roofing shingle.

Asphalts[1]		Granules
Coating Grade:		Gradation: -2.4 mm, +0.43 mm
Softening Point	82-104° C	Type: Igneous Rock or similar
Penetration @ 25° C	15-40 dmm	Mohs Hardness: >4
Flash Point	≥ 225° C	
Saturant Grade:		
Softening Point	48-60° C	
Penetration @ 25° C	50-150 dmm	
Flash Point	≥ 225° C	
Backsurfacing		**Mineral Fillers**
Gradation: ≥ 0.150 mm (plus No. 100 sieve)		Gradation: -0.150 mm (minus No. 150 sieve)
Type: silica sand, talc or limestone		Type: limestone, boiler slag, rock silica sand
Inorganic Fiber/Glassmats		**Organic Fiber/Felts**
Type: C or E glass		Type: Cellulose
Length: 6-32 mm		Length: 5 micrometer-5 mm
Diameter: 15-30 micrometer		Diameter: 2-5 micrometer

Other major product types can be readily conceptualized as variations of asphalt roofing shingles (see Fig. 1).

[1] Saturants and coating asphalts are both made from an asphalt flux by a process known as "blowing." During this process, air is bubbled through a large tank containing the hot flux. Heat and oxygen cause a chemical reaction which changes the characteristics of the asphalt. The process is monitored, and the "blowing" is stopped when the correct properties are produced.

Asphalt Saturated Felts/Glassmats, ply or base sheets: consist of the substrate and an asphalt impregnate or saturant. They contain no top or back asphalt coatings or inorganic surfacing material.

Mineral Surfaced Roll Roofing: conform to the same general composition as a shingle, but are manufactured in roll form rather than sheet form.

Modified Bitumen Membranes: conform to the same general composition, but contain a polymeric modifier in the asphalt coating such as Atactic Polypropylene (APP) or Styrene Butadiene Styrene block co-polymer (SBS). They may also contain more than one substrate and the substrates may be synthetic. Their basic form is a 10 m^2 roll.

The products are applied in a variety of configurations, but generally include at least one or more layers.

ABOUT ASPHALT IN ROOFING WASTE

In studying the physical properties of aged asphalt in roofing products, it is important to understand both the oxidative aging and stearic hardening aging mechanisms.

In multiple layer applications, the asphalt in the weather side (top exposed) layers undergoes predominantly oxidative age hardening. However, the majority of the product mass is in the unexposed lower layers, such as the headlap portion of shingles, saturated felts, interply membranes or interply built-up roofing asphalts, which age primarily via stearic hardening. The stearic hardening process has been demonstrated to be reversible by reheating and/or solubilizing. The general physical properties of reclaimed asphalt contained in commercially reclaimed roofing waste, ReACT's HMA™, has been evaluated to have a 70-82° C softening point and a penetration of 15-60 dmm.

ABOUT MINERAL FILLERS IN ROOFING WASTES

The asphalt roofing industry utilizes fine (minus 0.150 mm) inorganic mineral fillers such as limestone. These fillers are selectively specified by the roofing industry to stabilize the asphalt and retard aging. Their form does not change upon aging, and they remain encapsulated by asphalt through the reclamation process.

ABOUT FIBERS AND REINFORCING SUBSTRATES IN ROOFING WASTES

Reinforcing substrates of predominantly non-woven fiberglass or cellulose fibers are used to provide the carrier for the manufacturing process and the structural reinforcement on the roof. These reinforcing substrates are encapsulated by asphalt and, in the case of the cellulose fibers, also saturated with the saturant grade of asphalt (See Table 1). Their form and properties do not change significantly upon aging. Reclamation reduces the fiber length to less than 0.850 mm.

Review of existing hot-mix asphalt (HMA) additives and/or modifiers indicates each of the prime constituents of recycled asphalt roofing materials (RAR), by themselves, have been used functionally in a wide array of mix designs.

Each RAR constituent has a number of commercial equivalents currently being marketed for HMA pavement modification (Table 2).

TABLE 2--Commercial/functional equivalents to RAR.

RAR Component	Commercial/Functional Equivalent
Fibers	Minerals, Cellulose in SMA's, Polyester, & Polypropylene
Fillers	Carbon Black, Limestone, Hydrated Lime, Diatomaceous Earths
Air-Blown Asphalts	Gilsonites, Trinidad Lake Asphalt Propane precipitated asphalts

In the case of two well-known modifiers, Trinidad Natural

Asphalt™ [1] (TNA) and Arbocel® [2] (ARB), each consists of two components. TNA is composed of natural asphalt bitumen and fine mineral filler while ARB/Viatop® is a combination of high softening point asphalt and cellulose fibers [2, 3]. It is not surprising, then, that properly reclaimed asphalt roofing waste would exhibit enhanced performance properties in conventional HMA dense-graded compositions and in the "new" gap-graded, stone mastic asphalt (SMA) compositions now gaining acceptance in the United States [4, 5].

ABOUT RECYCLING ASPHALT ROOFING

The RAR selected was produced by ReClaim, Inc. in its Kearny, New Jersey facility. The proprietary process (five years in development) subjects the RAR to an array of size reduction, separation and classification steps resulting in a uniformly sized material. The particle size selected through extensive research and development, optimizes the pavement composition properties. Larger particles do not readily disperse, functioning much like aggregate. Particles sized too small result in the fibers functioning as fillers. Properly processed RAR promotes mix dispersion and homogeneity, allowing the prime components of fibers, hard asphalt and fillers to synergistically interact within the pavement composition.

Presently, ReACT's HMA™ is available at costs below domestic cellulose fibers, and the recycled products such as ground tire rubber, bottle glass or polyethylene wastes. Additionally, the cost to process ReACT's HMA™ in mixes is the same as RAP. Special process or handling requirements such as blenders or mixers, mills, blowers, and injections systems, are not necessary. ReACT's HMA™ is commercially available in quantities up to 50,000 tons per year.

OBJECTIVE

This study was to evaluate the efficacy of a commercially-available RAR, ReACT's HMA™, in a cross-

---

[1] Trinidad Natural Asphalt™, is the tradename of Petro Source Asphalt Products Company, Salt Lake City, Utah.

[2] Arbocel®, is the tradename of J. Rettenmaier & Söhne, Ellwangen-Holzmühle, Germany.

sectional array of conventional dense-graded pavement compositions and SMA compositions.

It was hypothesized that a controlled gradation RAR could replace a portion of the neat asphalt cement requirements while improving Marshall Stability, enhancing rut resistance, improving resistance to thermal and fatigue cracking, and allowing for increased asphalt binder contents, i.e. SMA concepts, without the need for additional binders, fillers or binder thickening agents such as polymers or cellulose fibers.

SCOPE

The experimental design included two concepts of incorporation into an HMA design: 1) the "**additive**" approach which processes the RAR as recycled asphalt pavement (RAP), and 2) the "**modifier**" approach which processes the RAR as an asphalt binder modifier, such as selected ground tire rubbers, TNA, and block co-polymers. The main focus was on the "additive" method since it offers the simplest means of field use, and requires no special equipment. Experiment designs were selected to include differing aggregate types, asphalt binders, and mix designs in use by Florida, New Jersey and Pennsylvania to assess overall effectiveness.

MATERIALS

The overall composition of the recycled roofing waste product, ReACT's HMA™, used in this study, compared to published specifications, is displayed below (Table 3).

TABLE 3--ReACT's HMA™ composition.

Property	Specified Range	Tested Result
Gradation	100% passing 0.850 mm	100% passing 0.850 mm
	100% retained 0.212 mm	≤ 2% passing 0.212 mm
Asphalt Content	30-40 weight %	30.7 weight %
Fiber Content	5-8 weight %	6.8 weight %
Filler Content	50-65 weight %	62.5 weight %
Specific Gravity	...	1.608-1.668

Properties were additionally determined on the de-asphalted and recovered components. Soxhlet extraction and rotary evaporator procedures were used, with toluene as the selected solvent. The results of the component recovery analysis are exhibited in Tables 4, 5 and 6.

TABLE 4--Recovered asphalt (30.7 wt. %).

Property	Result	Test Method
Softening Point	75.7° C	ASTM D 36[1]
Penetration @ 4° C	5.0 dmm	ASTM D 5[2]
@ 25° C	36.8 dmm	
Viscosity @ 60° C	6 824 P	ASTM D 2171[3]
Kinematic Viscosity @ 135° C	728 cSt	ASTM D 2170[4]

TABLE 5--Recovered mineral filler (62.5 wt %).

Property	Result	Test Method
Gradation, wt. % Passing		
0.850 mm (No. 20 sieve)[a]	100.0 %	
0.212 mm (No. 70 sieve)[a]	39.0 %	ASTM D 451/[5]
0.075 mm (No. 200 sieve)[a]	18.0 %	452[6]
Type	mixed limestone & crushed granules	Stereo microscope

[a] U.S.A. Standard Sieves - ASTM Specification E-11.

[1] ASTM Test Method for Softening Point of Bitumen (Ring-and-Ball Apparatus).

[2] ASTM Test Method for Penetration of Bituminous Materials.

[3] ASTM Test Method for Viscosity of Asphalts by Vacuum Capillary Viscometer.

[4] ASTM Test Method for Kinematic Viscosity of Asphalts (Bitumens).

[5] ASTM Method for Sieve Analysis of Granular Mineral Surfacing for Asphalt Roofing Products.

[6] ASTM Method for Sieve Analysis of nongranular Mineral Surfacing for Asphalt Roofing Products.

TABLE 6--Recovered fibers (6.8 wt %).

Property		Result	Test Method
Average Size:	length	0.5-4.5 mm	Stereo
	diameter	0.025-0.075 mm	Microscope
Cellulose/Fiberglass		90 wt % cellulose	ASTM D 2939[1]
Approximate Mix		10 wt % fiberglass	Para. 8

Neat Asphalts

The neat asphalt cements used in the mixture designs as classified by ASTM Specification for Viscosity-Graded Asphalt Cement for Use in Pavement Construction (D 3381), were from in-use and approved sources (Table 7).

TABLE 7--Approved asphalt cement sources per ASTM D 3381.

Design	Asphalt Cement	Source
Florida	AC-30	Coastal
New Jersey	AC-20	Chevron
Pennsylvania	AC-20	Sun Refining
PRI's Experimental	AC-20	Chevron
	AC-10	Shell

Aggregates

The aggregates used were common, in-use aggregates and from approved sources within the respective states. The following displays overall gradations of the separate aggregates used in the respective mixes (Table 8).

[1] ASTM Method of Testing Emulsified Bitumens Used as Protective Coatings.

TABLE 8--Gradations of aggregates.

Sieve Size Designation/Opening		Percentage Passing: Florida (S-1) #1	#2	#3	New Jersey (I-2) #1	#2	#3	#4	Pennsylvania (ID-2) #1	#2	#3
1	25.0 mm	...	...	...	100	100	100	100	...	100	...
3/4	19.0 mm	100	100	100	...	...	...	...	...	...	...
1/2	12.5 mm	91	100	100	35.6	100	100	100	100	45	...
3/8	9.5 mm	73	100	100	...	...	...	...	92	...	100
No. 4	4.75 mm	42	100	100	5.6	52.2	100	100	25	4	99.5
No. 8	2.36 mm	...	...	...	2.2	25.5	75	98	4	2	78
No. 10	2.00 mm	6	100	99	...	...	...	...	2	...	...
No. 16	1.18 mm	...	...	...	...	...	...	...	...	...	43
No. 30	0.600 mm	...	...	...	...	...	...	...	...	...	23
No. 40	0.425 mm	1	53	92	...	...	...	...	...	...	...
No. 50	0.30 mm	...	...	...	1	13.3	32.6	35	...	...	12
No. 80	0.180 mm	1	24	27	...	...	...	...	...	...	5
No.100	0.150 mm	...	...	...	...	...	...	...	...	...	3
No. 200	0.075 mm	0.9	7.7	8	0.5	4	14	1.6	...	...	...
Specific Gravity		2.350	2.475	2.615	2.89	2.89	2.89	2.64	2.710	2.602	2.710
Source		Florida DOT			Trap Rock Industries				Eastern Industries		

Aggregate Blends

The Florida aggregate blend is a standard "S-1" dense-graded design, using limestone aggregate (#1), limestone screenings (#2) and sand (#3). The typical mix design is 55 weight percent (#1), 25 weight percent (#2) and 20 weight percent (#3). For the experimental mix utilizing the RAR, the ratio was changed to 55 weight percent (#1), 20 weight percent (#2), 15 weight percent (#3) and 10 weight percent (RAR or #4).

The New Jersey aggregates were crushed trap rock supplied by Trap Rock Industries, and a washed sand supplied by Clayton Sand Co. The typical mix design for NJ DOT's I-2 base coarse design is 45.0 weight percent (#1), 15.0 weight percent (#2), 30.0 weight percent (#3) and 10.0 weight percent (#4). For the experimental mix utilizing the RAR, the percentages of #1, and #2 were kept the same and 5 weight percent RAR was substituted for an equivalent weight percent of #3 and #4. The overall gradation was within the NJ DOT specifications.

The Pennsylvania aggregates were #57 limestone (#1), #8 gravel (#2), and a washed sand (#3) supplied by Eastern Industries. The standard mix design for PA DOT (ID-2) wearing coarse is composed of 46.4 weight percent #3 and

47.2 weight percent #2 with 6.4 weight percent AC-20. For the experimental mix, 5 weight percent RAR was substituted for 3.5 weight percent #3 and 1.5 weight percent asphalt.

MIXTURE DESIGNS

The Marshall method of mix design, per the Asphalt Institute Manual Series No. 2 and ASTM Test Method for Resistance to Plastic Flow of Bituminous Mixtures Using Marshall Apparatus (D 1559), were used to prepare and evaluate all mixtures. The RAR was added directly to the heated aggregates, followed by the neat asphalt binder addition. Mixing and compaction temperatures were raised slightly to 145-163° C to accommodate for the increased viscosity of the resultant modified binder. A target air voids content of 4.0% was selected by varying asphalt binder contents. The level of neat asphalt cement was adjusted to achieve the targeted air voids content. A 50 blow compaction was used.

Five and ten percent RAR were targeted for direct addition to adjusted mix designs ("**additive method**"), based on previous experimental work. Usages over ten percent are possible, but require reduced neat asphalt contents and lower viscosity neat binders (and/or rejuvenating agents) which may not be readily available or may be more costly. Lower levels, below five weight percent, are feasible. At additive levels below five weight percent, the degree of neat asphalt savings and/or performance property improvement is diminished. To realize optimum performance properties at low levels of RAR use, pre-blending with the neat asphalt cement ("**modifier method**") is preferred, but requires additional on-site specialized binder mixing/blending equipment.

BINDER MODIFICATIONS

Seven various binder modifications were prepared to compare binders modified with RAR to known commercialized modifiers of both similar and dissimilar technologies. For the SBS modification, the neat asphalt was the same AC-30 used in the other experimental design elements. (**NOTE**: Typically, a lower viscosity grade of neat asphalt would be recommended by the SBS supplier to compensate for the resultant viscosity increase.)

Levels of addition for each modifier were those generally recommended by the specific technology (Table 9).

TABLE 9--Comparison of modifiers.

Symbol-Modifier Type		% Added	Comments
PE	Polyethylene	5.5	Novophalt[R][1] Blend supplied to Florida Dept. of Transportation for additional testing
SBS	Styrene-butadiene-styrene block co-polymer	4.0	with Vector 2411 SBS supplied by Dexco
ARB	Arbocel Pellets/Viatop	8.0	4% asphalt, 4% fibers equal to 6 lbs. fiber/ ton of mix supplied by ScanRoad
RAR	Recycled roofing waste	28.4	9.9% asphalt, 2.3% fiber & 16.2% filler
DE	Diatomaceous earth	10.7	Concept used in Europe; Celite 282 supplied by Manville
TNA	Trinidad Natural Asphalt	25.0	12.5% asphalt, 12.5% filler; supplied by Petro Source
CF	Cellulose fiber	4.0	CS 31500 supplied by Custom Fiber Int'l
CTRL	None	0.0	Control

All blending procedures were equilibrated to 60 minutes at 175° C to maintain the same heat history on the blend, including the control. The only exceptions were:

PE was milled at 170° C using a continuous, high-shear mill with the rotor-stator gap pre-set at 0.254 mm. Two passes were made. The total extra heat history was estimated at 10 minutes.

SBS was milled at 185-190° C using a high-shear homogenizer, batch system, with a rotor-stator clearance of 4.06 mm. The estimated extra heat history was 90 minutes.

---

[1] Novophalt[R] is a registered trademark of Novophalt America, Inc., Sterling, Virginia.

EVALUATIONS

For the "additive method," a combination of testing protocols were utilized for the design mixes, including Marshall Properties, Indirect Tensile Strength [6], and Georgia Loaded Wheel Tester [7, 6].

For the "modifier method," conventional and newly-proposed Strategic Highway Research Program (SHRP) protocols [8] were followed to investigate and compare various modified binder properties.

RESULTS - "ADDITIVE" METHOD

Key Marshall Specimen properties for the Florida S-1 design [6], the New Jersey I-2 base coarse mix design [9], and the Pennsylvania Section 401 ID-2 wearing mix design [10] are displayed in Tables 10, 11, and 12, respectively.

TABLE 10--Marshall Specimen properties for Florida S-1.

Marshall Property	Mix w/ 10% RAR	Control (1)	Control (2)
Total wt. % Asphalt	7.00	7.00	6.0
Sp Gr of Compacted Mix	2.180	2.207	2.184
Max. Gravity	2.287	2.300	2.331
Density, kg/m^3	2 187	2 205	2 182
Volume of Effective AC	7.7	11.5	9.2
% Air Voids	4.7	4.0	6.3
Voids in Mineral Aggregate	12.4	15.5	15.5
% Voids Filled	62.1	74.2	59.4
Avg. Stability, N	18 770	7 459	8 064
Avg. Flow, mm	3.23	2.24	1.78

TABLE 11--Marshall Specimen properties for New Jersey I-2.

Marshall Property	Mix w/ 10% RAR	Typical Control (1)
Total Wt. % Asphalt	6.3	4.5
Sp.Gr. of Compacted Mix	2.514	2.598
Max. Gravity	2.630	2.690
Density, kg/m^3	2 192	2 613
% Air Voids	4.4	3.4
Voids in Mineral Aggregate	20.3	14.8
% Voids Filled	78.4	77.0
Avg. Stability, N	22 567	13 179
Avg. Flow, mm	3.3	3.3

TABLE 12--Marshall Specimen properties for Pennsylvania ID-2.

Marshall Property	Mix w/ 5% RAR	Typical Control (1)
Total Wt. % Asphalt	6.5	6.4
Sp.Gr. of Compacted Mix	2.341	2.339
Max. Gravity	2.439	2.439
Volume of Effective AC	...	5.9
% Air Voids	4.0	4.1
Voids in Mineral Aggregate	18.9	17.5
% Voids Filled	78.6	76.6
Avg. Stability, N	12 668	10 528
Avg. Flow, mm	2.60	2.80

Selected S-1 Florida mix designs containing ten weight percent RAR were also evaluated for Indirect Tensile Strengths per ASTM Method for Indirect Tension Test for Resilient Modulus of Bituminous Mixtures (D 4123). The results are displayed below (Table 13) [6].

TABLE 13--Indirect tensile strengths - Florida S-1.

		INDIRECT TENSILE STRENGTH @ 25° C		
Total Asphalt Binder	Neat AC-20	Ultimate Load (kg)	Tensile Strength, kg $\bar{x}$	Std. Dev.
7.0	3.5	1792.9	1755.9	77.9
8.5	5.0	1217.3	1187.1	102.7
10.0	6.5	856.2	817.6	57.2
6.5 (Control)	6.5	935.2	901.7	31.7

The Control and the experimental mix containing ten percent RAR and with a total asphalt cement content of 7.0 weight percent were evaluated for rut resistance in accordance with Georgia DOT Loaded Wheel Tester procedures [7, 6]. The results are graphed (Fig. 2) [6]. The following test conditions were observed: 689.5 KPa, 38.9-40.9° C for all samples tested.

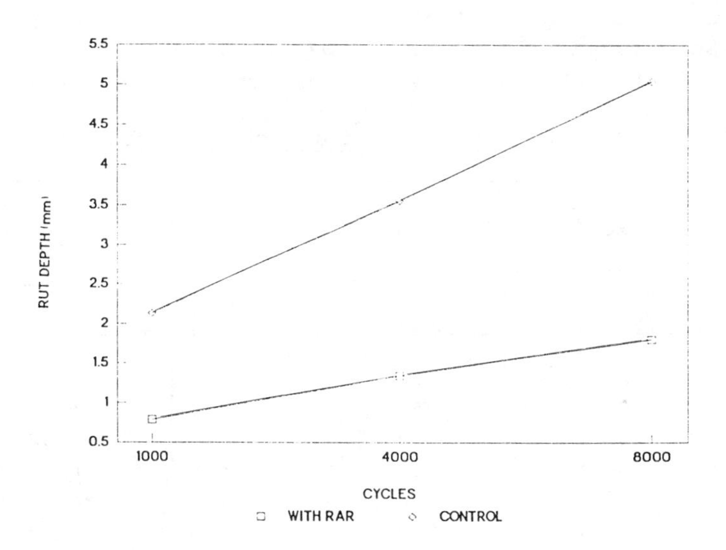

FIG. 2--Graph showing higher stability and better rut resistance of the experimental mixes.

RESULTS - "MODIFIER" METHOD

A "drain-down" test [11] was performed on laboratory-prepared mixes which approximated a typical SMA design (Table 14). Aggregate materials used were Florida DOT approved and supplied by Couch Construction Company, Tampa, Florida.

TABLE 14--SMA mix design blend.

Material Description	Gradation	Wt. %
Experimental Modified Asphalt Binder	N/A	7.5
Coarse Aggregate	- 12.7 mm, + 4.75 mm	57.0
Medium Aggregate	- 4.75 mm, + 2.36 mm	20.0
Crushed Sand	...	13.0
Fine Fillers	- 0.075 mm	10.0

One kilogram of each mix was prepared at 170-175° C; 500 gram aliquots of non-compacted mix were conditioned for one hour at 170° C in a glass beaker. After conditioning, the mix was poured out. The residual binder adhering to the beaker was weighed. The results, expressed as a weight percent of the total binder, are displayed in Table 15.

TABLE 15-Modified binder drain-down percents.

Modified Binder				
Asphalt Cement		Modifier		Drain-Down
Type	Wt. %	Type	Wt. %	Wt. %
AC-30	100	...	...	5.64
AC-10	71.4	RAR	28.6	2.29
AC-30	96	SBS	4	3.52
AC-30	94.5	PE	5.5	1.32
AC-30	75	TNA	25	7.95
AC-30	87.5	ARB	12.5	0.40
AC-30	96	CF	4	0.75

The SHRP Test Method B-002, Bending Beam Rheometer (BBR) [8], was used to provide insight into the thermal and fatigue cracking of the modified binder after processing. The BBR test was conducted before and after simulated process aging (TFOT) at: -15°C, 100 gm load for 240 seconds, on a specimen beam 125 mm long, 12.5 mm wide and 6 mm thick. The results are shown below (Table 16).

TABLE 16--Bending Beam Rheometer: m Value.

Modified Binder	Effective AC Wt. %	m Value	
		Before Aging	After Aging
Control - AC-30	100	0.37	0.35
AC-30 (94.5%)/PE (5.5%)	94.5	0.36	0.35
AC-30 (96%)/SBS (4%)	96	0.33	0.32
AC-30 (92%)/ARB (8%)	96	0.35	0.32
AC-30 (71.4%)/RAR (28.6%)	89	0.37	0.32
AC-30 (89.3%)/DE (10.7%)	89.3	0.40	0.35
AC-30 (75%)/TNA (25%)	87.5	0.37	0.33
AC-30 (96%)/CF (4%)	96	0.37	0.34

The m value at 60 seconds, was comparable between modifiers before or after simulated process aging (TFOT).

Figure 3 displays the viscosity of the samples defined in Table 9, in centipoise (cps) at 135°C, as determined by ASTM Method for Viscosity Determinations of Unfilled Asphalts Using the Brookfield Thermocel Apparatus (D 4402), before and after TFOT aging.

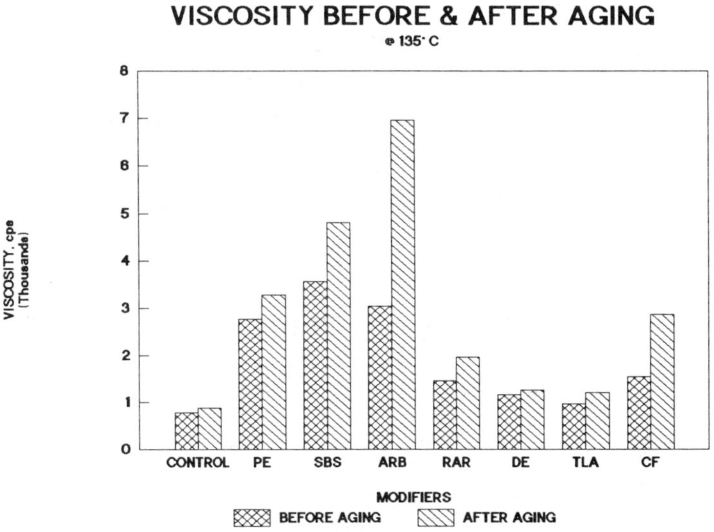

FIG. 3--Viscosity before and after TFOT Aging.

Figure 4 displays the conventional softening point data of the same samples, before and after TFOT aging.

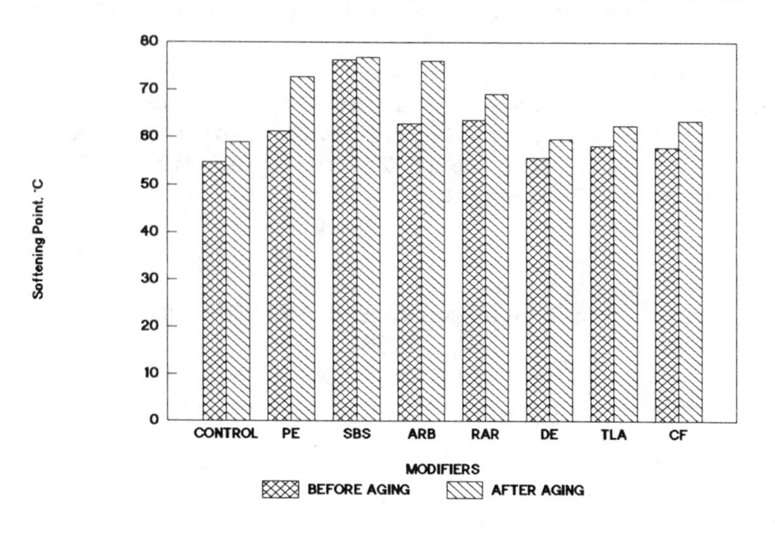

FIG. 4--Softening point data before and after TFOT aging.

CONCLUSIONS

Additive Method: Conventional Dense-Graded HMA Designs

1. Commercially-prepared RAR can be used to replace a portion of the neat asphalt binder and aggregate resulting in mixes with equivalent and/or improved properties as measured by Marshall protocols.

2. The use of RAR in conventional dense-graded mix designs has a wide application and yields similar improvements for mixes using different aggregate types, aggregate gradations, asphalt binder grades, and overall designs.

3. The incorporation of RAR benefits the mix design in a synergist manner, with all three prime components acting together to improve high temperature pavement deformation properties, such as rut resistance.

4. The RAR can be used in conventional equipment ("additive" method) similar to processing RAP. The use of specialized equipment and/or processes is not required.

5. HMA pavement designs requiring high stabilities, i.e. rut resistance, for high-traffic or load-bearing areas, can be readily achieved with the use of RAR.

Modifier Method: SMA Gap-Graded HMA Designs

6. The RAR exhibited excellent drain-down resistance when compared to in-use commercial modifiers added for the same purpose.

7. RAR was easier to add and blend than pelletized asphalt cellulose fiber, cellulose fiber, polyethylene, SBS, and TNA.

8. Viscosities of RAR blends are increased. Mixing and/or binder temperatures will require a 5-10° C increase to achieve equal viscosities of control binders.

9. Creep Stiffness determined by the BBR indicates RAR blends, are more resistant to deformities at low temperatures. Low-temperature thermal cracking would not be expected based on comparisons to other known good-performing modifiers.

SUMMARY

Recycled asphalt roofing waste, ReACT's HMA™, as commercially supplied by ReClaim, Inc., Tampa, Florida, is a multi-functional HMA additive/modifier that can be used in both conventional dense-graded and SMA mix designs to replace neat asphalt and increase overall binder contents.

This study suggests the use of recycled roofing waste will provide pavement designs with increased stabilities, reduced tendencies to rut, and improved high-temperature properties. Further, ReACT's HMA™ was extremely easy to use. Blending and/or mixing was accomplished within industry-accepted parameters.

Evaluations with ReACT's HMA™ indicated it would be an excellent candidate for economical SMA mixes. Proper use would increase overall binder contents, reduce neat asphalt requirements, eliminate the need for additional modifiers, and would only require conventional processing equipment and parameters.

The use of asphalt roofing waste in HMA's recycles a current waste source into higher performing HMA designs in a cost-effective manner.

ACKNOWLEDGEMENTS

PRI Asphalt Technologies, Inc. and the author wish to gratefully acknowledge ReClaim, Inc. for providing funding and support for this work.

The author also gratefully acknowledges the inter-laboratory support and data sharing provided by Messrs. Ken Murphy and Phil Eunice of the Florida DOT; Jeff Frantz, Technical Director, Eastern Industries; and Wayne Byard, Technical Director, Trap Rock Industries.

Additionally, the author is thankful to David R. Jones, IV, Ph.D., SHRP Research Chemist and Vyt Puzinouskas, for reviewing and sharing their comments on this paper.

REFERENCES

[1] Snyder, R. D., Letter: "Best estimates for '92 shipments," Asphalt Roofing Manufacturer's Assn., 1992.

[2] "Arbocel^R Asphalt," product pamphlet, J. Rettenmaier & Söhne, Ellwangen-Holzmühle, Germany.

[3] "VIATOP^R The latest development in granulated Arbocel^R fibers for the road building of the future," product pamphlet, J. Rettenmaier & Söhne, Ellwangen-Holzmühle, Germany.

[4] Warren, J. M., "SMA Comes to the USA," Hot Mix Asphalt Technology, pp. 5-9, Vol. 6, No. 2, National Asphalt Pavement Assn., Lanham, Maryland, 1991.

[5] "European Cinderella Mix Introduced in U.S. Tests," Asphalt Contractor, pp. 50-51, 68, September-October 1991, Asphalt Contractor Publishing Co., Independence, Missouri, 1991.

[6] Murphy, K. H., Laboratory Report - "ReACT's HMA™ for Use in HMA," State Materials Office, Florida Department of Transportation, Gainesville, Florida, 1992.

[7] West, R. C., et al., "Evaluation of the Loaded Wheel Tester," Research Report, FL/DOT/SMO/91-39/, State Materials Office, Florida Department of Transportation, Gainesville, Florida, 1991.

[8] Technology Information Sheet: "SHRP Test Method B-002, Bending Beam Rheometer," Strategic Highway Research Program, Washington D.C., 1992.

[9] Byard, W., Laboratory Report - "I-2 Drum Mix with ReACT's HMA™," Trap Rock Industries, Kingston, New Jersey, 1992.

[10] Frantz, J. J., Laboratory Report - "Bituminous Mix Design Evaluation of Recycled Asphalt Roofing Material," Eastern Industries, Center Valley, Pennsylvania, 1992.

[11] Technotes: "Split Mastic Asphalt Manual and Effect of Additives on Rutting Resistance of Split Mastic Asphalt," Elf Asphalt, Raleigh, North Carolina, 1992.

Petroleum Contaminated Soils

Namunu J. Meegoda[1], Robert T. Mueller[2], De-Rong Huang[3], Bonnie H. DuBose[3], Yaoqing Chen[3] and Kuen-Yuan Chuang[3]

PETROLEUM CONTAMINATED SOILS IN HOT MIX ASPHALT CONCRETE - AN OVERVIEW

REFERENCE: Meegoda, N. J., Mueller, R. T., Huang, D. R., DuBose, B. H., Chen, Y., and Chuang, K. Y., **"Petroleum Contaminated Soils in Hot Mix Asphalt Concrete - An Overview,"** Use of Waste Materials in Hot Mix Asphalt, ASTM STP 1193, H. Fred Waller, Ed., American Society for Testing and Materials, Philadelphia, 1993.

ABSTRACT: Since 1985, petroleum contaminated soils (PCSs) have been used in the production of hot mix asphalt (HMA) and PCS in HMA has become an attractive recycling option for PCS. When PCSs are used in the production of hot mix asphalt concrete three beneficial actions occur: incineration, dilution and solidification. In this paper, based on laboratory and field studies, the performance and suitability of construction of hot mix asphalt concrete with petroleum contaminated soils is discussed. Production of hot mix asphalts with several different contaminated soils is first investigated. Then the following engineering issues are discussed based on the laboratory and field data for hot mix asphalts with petroleum contaminated soils: 1) the impact with respect to strength and durability of asphalt concrete, and 2) change in secondary performance parameters of asphalt concrete such as hydraulic conductivity with the inclusion of petroleum contaminated soils.

KEY WORDS: petroleum contaminated soils, hot mix asphalt, solid waste, recycling, stability, durability, freeze-thaw, wet-dry, hydraulic conductivity

---

Petroleum contaminated soil (PCS) is generated from leaking Underground Storage Tanks (USTs), including piping connected to USTs. During 1950s and 1960s, the construction of many gasoline stations, chemical manufacturing and processing facilities led to the installation of millions of USTs. Several million USTs in the United States contain petroleum products. Tens of thousands of these USTs, including their piping, are currently leaking [1]. The USEPA estimates that there are more than 400,000 leaking USTs with petroleum hydrocarbons. Many more are expected to leak in the future. Most states vigorously encourage the removal of all tanks after 25 years of service. It is estimated that on average removal of a leaking tank generates 50 to 80 cubic yards of contaminated soil. Since groundwater is the major source of drinking water, federal legislation seeks to safeguard our nation's ground water resources. Congress responded to the problem of leaking USTs and subsequent groundwater contamination by adding Subtitle I to the Resource Conservation and Recovery Act in 1984. The federal statues and statues from different states require the removal of such leaking USTs to prevent further contamination. Gasoline, diesel, and fuel oil are the most common petroleum hydrocarbons used in USTs and consequently are the ones that most likely will leak from USTs. Each of these products is a complex mixture of

---

[1]Associate Professor, Department of Civil and Environmental Engineering, New Jersey Institute of Technology, Newark, NJ 07102

[2]Research Scientist, Division of Science and Research, State of New Jersey Department of Environmental Protection and Energy, CN 409, Trenton, NJ 08625

[3]Graduate Student, Department of Civil and Environmental Engineering, New Jersey Institute of Technology, Newark, NJ 07102

organic compounds with specific physical and chemical properties and behavior when in contact with subsurface soil and water. The major chemical components of petroleum hydrocarbons are alkanes, cycloalkanes, and aromatics. The physical and chemical characteristics of each fuel depend on the type (i.e., gasoline, diesel, and fuel oil) and the source from which it was obtained (i.e., source of crude). Gasoline is the most volatile of the three, and major chemical components include branched chain paraffins, cycloparaffins and aromatics. Diesel is the No. 2 fuel oil. It is primarily composed of unbranched paraffins with flash points between 43° and 88°C. Fuel oils are chemical mixtures having flash points greater than 37°C. Compositions of fuel oil vary much more than that of gasoline and diesel.

Petroleum contaminated soils consist of mixture of gravels, sands, silts and clays with petroleum products. A small portion of light petroleum product mixed with asphalt merely produces an asphalt of slightly different specification or characterization. Therefore, it is viewed by many that a small increase in the quantity of light petroleum substances would not damage the HMA. This is the basis for the theory that contaminated soils can be used as a binder in asphalt concrete paving. However, the inclusion of soils in HMA requires an in-depth study.

The hot mix asphalt concrete used in asphalt roads consists of a combination of aggregates blended and uniformly mixed, coated with asphalt cement, and compacted. The aggregates in HMA consist of (a) course aggregates or gravel with size as large as 1.5" to US sieve #4, (b) fine aggregate or sand with sizes passing US sieve #4 and retained on US sieve #200, (c) mineral filler such as crush stone dust or lime passing US sieve #200, and (d) asphalt cement. A typical HMA composition consists of 55% coarse aggregate, 40% fine aggregate, and 5% mineral filler, and 5.5% asphalt cement. Asphalt cement is obtained by distillation of petroleum crude. The asphalt cements obtained from refineries are classified as AC-2.5, AC-5, AC-10, AC-20, AC-30, or AC-40 based on their viscosities. To obtain sufficient fluidity of asphalt cement for proper mixing and compaction, both the aggregate and asphalt cement are heated before mixing; hence it is called hot mix asphalt (HMA) concrete. Five to ten percent industrial waste products such as recycled asphalt pavements, tire rubber, glass, MSW ash, roofing shingles, polythene waste and petroleum contaminated soils [2,3] can be added to HMA without sacrificing the strength and performance of HMA. If as a rule 5% of waste products are included in all HMA produced, then based on 1988 USDOT estimate [4] 25 million tons of industrial waste can be recycled and consumed annually by the US Asphalt Industry. These waste products can be added to the HMA by either replacing the mineral filler or proportionately reducing the amount of virgin material in the original mix. Therefore, there is a major emphasis at all levels of governments in US to include waste products in HMA.

When PCSs are added to HMA, three beneficial actions occur; incineration, dilution and solidification. Part of the petroleum can be used as a fuel in the dryer as it is burned during the production of asphalt concrete. Thus a majority of the contaminants are beneficially eliminated and are used to reduce the fuel cost. There is spreading and dilution as only a small fraction (5% - 15%) f petroleum contaminated soil is added to virgin aggregates during the production of asphalt concrete. Because the asphalt cement acts as a binder in asphalt concrete, the remaining diluted contaminants are solidified and stabilized in the final asphalt concrete matrix.

Strength or stability, durability, and workability are the primary factors in a Hot Asphalt Mix Design [5]. The secondary factors are flexibility, permeability, fatigue resistance, skid resistance, and stripping action [5]. Since the HMA produced with PCS is used for paving operations, it should satisfy all the requirements specified by the appropriate federal, state or local agencies. Beside the above, solidified and stabilized petroleum products in the asphalt matrix should not leach (or come out) and should not cause any deterioration of air quality during the production of HMA with PCSs.

A laboratory and field research project, funded by the New Jersey Department of Environmental Protection and Energy (NJDEPE), was initiated in 1990 to investigate the feasibility of using PCSs in HMA. During the past two years laboratory and field studies were conducted to evaluate the use of PCSs as aggregate replacement in the production of hot mix asphalt concrete [6,7]. In this laboratory and field investigation petroleum contaminated soils were added to HMA, and the resulting asphalt concrete mixes were tested for strength, durability, permeability, and leachability. During the field production of HMA with PCSs emission of volatile organic carbon compounds (VOCs) was also monitored. In this paper the strength, durability, and permeability test results of the project are discussed. The emission of VOCs, leachability of organic compounds to air and water, and economic considerations of the process are not discussed in this paper.

EXPERIMENTAL PROCEDURE AND TEST RESULTS

Soil Classification

Six contaminated soils provided by NJDEPE from sites identified as containing soils with less than 1% total petroleum hydrocarbons (TPH) were selected for testing and for characterization. The soils were selected in such a manner to cover most of the soil types that occur in nature (gravel, sand, silt and clay, see ASTM D2487-85). Therefore, the proposed process will be a generic one for most soils and it will not be site specific. First three soils selected for testing were contaminated with heating oil and the latter three were contaminated with gasoline. One gasoline contaminated soil (soil #5) used in the lab study and two other heating oil contaminated soils were used in the field study. The degree of contamination for oil contaminated samples were determined by the Soxhlet oil and grease extraction method (USPHS standard method for the analysis of water and waste water). Tables 1 and 2 show the soil classification data, and type and amount of contaminants in each soil.

TABLE 1-- Data on Six Contaminated Soils from NJ used in the Lab Tests.

	Soil #1	Soil #2	Soil #3	Soil #4	Soil #5	Soil #6
Soil Classification	Well Graded Sand	Clayey Silt	Silty Sand	Poorly Graded Sand	Silty Clay	Poorly Graded Sand with Silt
In-situ Moisture Content (%)	7.3	14.3	24.7	14.4	19.6	10.1
Type & Amount of Contamination	1,100 ppm Heating Oil	1,200 ppm Heating Oil	6,600 ppm Heating Oil	25 ppm Gasoline	1500 ppm Gasoline	330 ppm Gasoline

TABLE 2-- Data on Three Contaminated Soils Used in the Field Tests.

	Soil #5	Soil #7	Soil #8
Soil Classification	Silty Clay	Silty Sand	Silty Sand
Type & Amount of Contamination	1,200 ppm Gasoline	2,000 ppm Heating Oil	3,200 ppm Heating Oil

Marshall Stability Test

All the asphalt concrete samples with PCSs were compacted in the laboratory for NJ I-3 mix. A control was designed and tested for comparison. The mix design for all the soils were performed based on the sieve analysis data. The optimum percentages that may be used in the production of NJ I-3 mix for the above six soils were as follows: Soil #1 - 35%; Soil #2 - 10%; Soil #3 - 20%; Soil #4 - 15%; Soil #5 - 10%; and Soil #6 - 15% [6]. The above percentages are much higher than the current practice of adding 5% PCSs to HMA [2]. Once the maximum amount of PCS that may be added to HMA was determined, the suitability of such an addition was evaluated. The Marshall Stability test was performed using ASTM D 1559 - 82. Based on the NJ requirements for Marshall strength, bulk density, air voids, VMA, and flow, the optimum asphalt content for each mix was selected. All the laboratory tests were performed with AC-20 asphalt cement. The optimum asphalt cement contents for the six contaminated soils are listed in Table 3. Then corresponding to this optimum asphalt content, the following were selected for the control as well as for HMA made with each soil type, using the average of three specimens, and reported in Table 3; Dry density, Marshall Stability, Air Voids, VMA and Flow values. Table 3 also shows the New Jersey specification for high traffic volume I-3 mix.

TABLE 3-- Optimum Properties of Asphalt Concrete with PCSs for NJ I-3 Mix

Asphalt Concrete Property	Allowable for NJ I-3 Mix	Control	Soil #1	Soil #2	Soil #3	Soil #4	Soil #5	Soil #6
Strength (N)	>8006	8006	8228	8450	10229	8450	8317	10452
Flow (0.25 mm)	>6.0	4.0	11.0	8.0	7.5	3.5	6.5	7.7
Air Voids (%)	3.0 - 8.0	7.0	7.5	3.0	5.7	8.0	4.0	3.4
VMA (%)	> 13.0	18.0	17.8	14.0	16.8	18.0	14.7	14.2
Density (kN/m^3)	N/A	24.3	24.8	24.5	24.1	23.4	24.6	24.5
Optimum Asphalt Content (%)	4 - 8	5.0	4.5	4.5	5.0	4.5	4.5	4.5

N/A: not available

Field tests were conducted at Continental Paving, Inc., Londonderry, New Hampshire to evaluate the applicability of adding PCS to HMA. This asphalt plant is a modified drum mix plant. The original plant design, with a 2.4m diameter and 13.7m length drum, and the initial modification to the plant to accommodate PCSs were designed and fabricated by the Astec Industries, Chattanooga, TN. The subsequent modifications were designed and fabricated by the owner to enhance the quality of air emitted from the stack. The latter modification consisted of attaching an additional drum that served as a soil desorber to thermally desorb the contaminated soil. The extended segment (a 2.7 m diameter and 9.1 m long drum to desorb the soil) was attached to the front of the original drum. When the heated soil enters the older section of the drum the cold aggregates were added. The plant is also capable of adding contaminated soil with cold aggregate. This option made it act as an ordinary hot mix asphalt plant. The plant is capable of producing 300 tons of HMA per hour and the best production rate was achived at 200 tons per hour. Two truck loads of soil #5 (52 tons) were sent to this plant from New Jersey in April 1991. The soil was processed before the field visit to remove large particles and debris (which resulted in 45 tons to produce HMA). For description the tests were divided into different groups based on the mix designs and are shown in Table 4.

TABLE 4-- Optimum Aggregate/Soil Percentages for Field Mixes

Aggregate Type	Test #1	Test #2	Test #3	Test 4	Test 5
3/4" Aggregate	30.0	27.9	50.0	27.9	0.0
1/2" Aggregates	16.5	4.1	0.0	4.1	0.0
3/8" Aggregates	20.5	27.3	0.0	27.3	39.6
River Sand	24.0	20.5	0.0	20.5	43.4
PCS Type	Soil #5	Soil #7	Soil #8	Soil #8	Soil #8
PCS Amount	9.0	20.2	50.0	20.2	17.0

Test #1 produced an NJ I-3 mix. The design asphalt content for test #1 was 4.8% and the actual asphalt content, from extraction test, was 4.5%. Fig. 1 shows the extraction test results. Test #2: produced a New Hampshire 3/4" base mix. The design asphalt content for test #2 was 4.9% and the actual asphalt content, from extraction test, was 5.17%. Test #3 produced a filling product for sub bases with a design asphalt content of 2.0%. The soil was fed at a rate of 50.0 tons per hour and when the soil came out of the extended dryer that used for thermal desorption, its temperature was 343°C. Since this product was used as filling material with low asphalt contents it was difficult to produce Marshall specimens. Test #4 produced a New Hampshire 3/4" base mix. The design asphalt content was 4.9% and the actual asphalt content, from extraction test, was 5.17%. The soil was fed at a rate of 38.0 tons per hour and when the soil came out of the extended dryer its temperature was 400°C. Test #5 produced a New Hampshire 3/8" surface mix. The design asphalt content was 6.1% and the actual asphalt content, from extraction test, was 6.38%. The soil was fed at a rate of 25.5 tons per hour and when the soil came out of the extended dryer its temperature was 454°C.

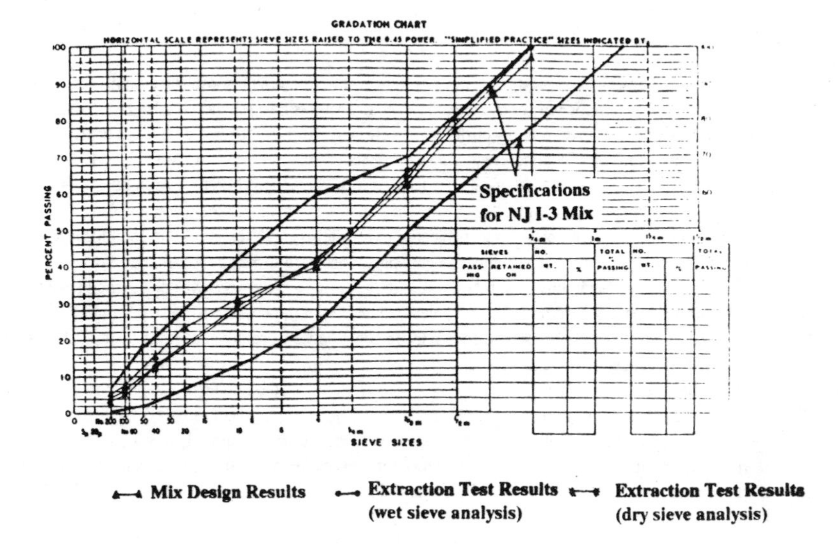

FIG. 1-- The NJ I-3 Specification, Design Grain Size Distribution, and Extraction Test Results from the Field Test.

Test #1 and #2 took nearly twenty-five minutes to complete and each test consumed ten tons of PCS. Since these two tests were performed for field demonstration, the HMA produced was recycled. The HMA with PCS #8 was used to pave roads. The test # 3 lasted two hours and produced 200 tons of sub base material. The test # 4 lasted three hours and produced 450 tons of 3/4" base mix. The test # 5 also lasted for three hours and produced 450 tons of 3/8" surface mix. For each test variation, as indicated, the aggregates and asphalt cement were added to the soil in the drum. The hot mix asphalt with PCS was allowed to progressed down to the drum, where it was mixed, until it reached a conveyer belt leading up to silo #1. After fifteen minutes, when sufficient amount of asphalt concrete was produced, the conveyer belt was moved to silo #2. A dump truck was positioned below the silo #2 and filled it with the hot mix asphalt concrete. Nine specimens (three for stability, three for durability and three for permeability) each weighing approximately 1250g were taken and compacted into Marshall samples and were brought back to the NJIT (New Jersey Institute of Technology) lab for further testing. Table 5 shows the Marshall test results of specimens from each test. Test #1 and #2 were produced with AC-20 asphalt while Test #3, #4 and #5 were produced with AC-10 asphalt. Therefore, tests #4 and #5 produced mixes with low stability values.

TABLE 5-- The Marshall Stability Test Results for Field Specimens.

Test #	Stability [N]	Flow [0.25 mm]	Air Voids [%}	Density [kN/m^3]	VMA [%]
1	11697	3.0	5.3	23.7	16.4
2	7872	5.8	6.5	23.4	17.4
3	N. D.	N. D.	N. D.	N. D.	N. D.
4	7263	10.0	8.5	22.7	20.1
5	5177	6.0	7.8	23.0	19.3

N. D: not done

Durability of HMA with PCSs

ASTM D 4867 describes the test procedure on how to find the effect of moisture on asphalt concrete mixtures, a factor that is very important for the durability of concrete. It has a section on freeze-thaw conditioning of a mixture. However, the freeze-thaw and wet-dry tests were only subjected to one cycle each of freeze-thaw or wet-dry. Since the real world asphalt concrete is subjected to several freeze-thaw and wet-dry cycles under service conditions before it is removed for resurfacing, it was decided to subject the HMA to several durability cycles. Therefore, eighteen HMA specimens with PCS #3 were compacted to evaluate the durability of HMA subjected to several durability cycles. Two specimens were subjected to tensile strength test to obtain the tensile strength without subjecting to environmental conditioning. Eight specimens were subjected to freeze-thaw cycles. After one cycle of freeze-thaw two specimens were taken out and the tensile strength test was performed to obtain the tensile strength and the tensile strength ratio (TSR) value. The rest of the specimens were continued subjecting to freeze-thaw cycles and after a total of three cycles two more specimense were taken out and tested for tensile strength. The rest of the specimens were tested after seven and fourteen cycles respectively. The remaining eight specimens were subjected to a series of wet-dry cycles similar to that of freeze-thaw tests. The percentage swell and percentage change in weight of specimens were also calculated before the tensile strength test. The above procedure was also adopted for the eighteen specimens made from the control mix. The results of these test series are plotted in Fig. 2, 3, 4 and 5. Fig. 2 shows the variation of tensile strength ratio, and percentage swell with the number of freeze-thaw cycles, for HMA with PCS #3 and Fig. 3 shows the same for the control mix. Fig. 4 and 5 show the similar data for wet-dry tests.

Testing for 1, 3, 7, and 14 cycles continued for HMA with PCS #3 and for the control mix with each cycle taking approximately 48 hours to complete. Upon completing 14 cycles of freeze-thaw and wet-dry, data was collected and graphically displayed. The performance of the control mix was quite similar to that of HMA with PCS #3. It can be concluded from these test results that as the temperature of the specimen drops during the freeze cycle and the asphalt concrete contracts and become brittle, creating micro cracks on the surface of the specimen that provides entry points for water. Water inside the specimen caused moisture damage due to stripping and volume expansions during subsequent freeze cycles. With the increase in number of freeze-thaw cycles cracks got larger letting more water into the specimen eventually leading to the failure of the specimen. The data from cyclic freeze-thaw test indicated that the percentage swell increased rapidly during the first cycle and then gradually reached a maximum before the specimen failed. It is believed that the specimen reached its maximum percentage swell when it was totally saturated and stripping occurred before complete failure. The tensile strength ratio also declined rapidly during the first cycle and then began to level off to a zero strength after 14 cycles. It seems that there is a correlation between the percentage swell and the tensile strength ratio. From the data, it shows that most of the strength is lost during the first cycle, and hence there was no need to test beyond one freeze-thaw cycle as suggested by the ASTM. The cyclic wet-dry test also indicated that the first cycle was the point where attention must be focused. For the wet-dry test, the tensile strength ratio declined during the first cycle and increased thereafter. This is believed to be due to oxidation of asphalt during the drying cycle. During the repeated drying and wetting of the specimen asphalt cement may have oxidized to form a brittle mix with higher tensile strengths. Therefore, it was concluded that the first wet-dry cycle should yield the critical conditions.

Therefore, the durabilities of hardened HMA concrete with all six PCSs were determined using the wet & dry and freeze & thaw tests (ASTM D 4867-88). The TSR value was used to evaluate the durability of each mixture. Table 6 shows the TSR values of wet-dry and freeze-thaw tests for the control and for the HMA made with all six soils.

The durabilities of field compacted and hardened HMA concrete with PCSs were also determined using the wet & dry and freeze & thaw tests (ASTM D 4867-88). The Tensile Strength Ratio (TSR) was used to evaluate the durability of each mixture. Table 7 shows the TSR values of wet-dry and freeze-thaw tests for the HMA made five field tests.

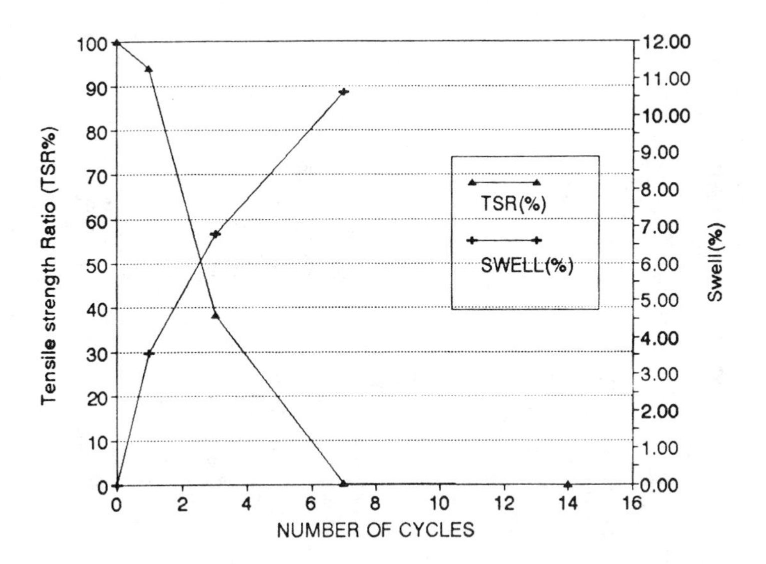

FIG. 2-- The extended Freeze-thaw Test Results for HMA with PCS #3.

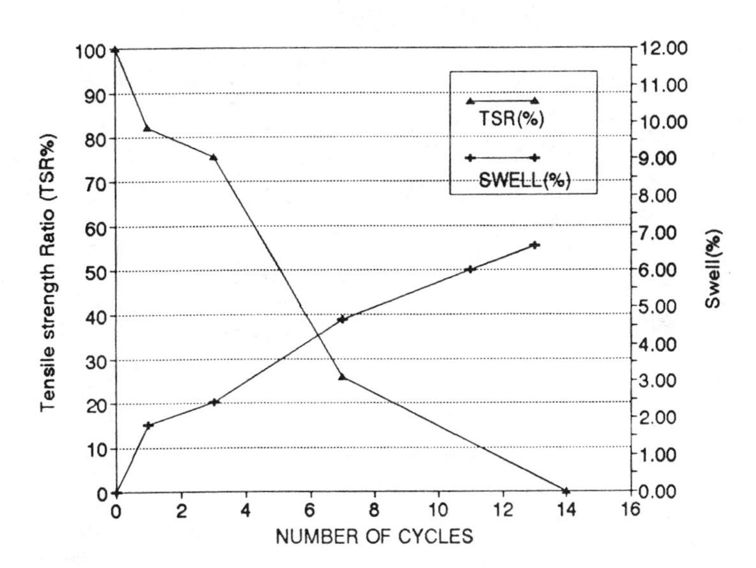

FIG. 3-- The extended Freeze-thaw Test Results for the control mix.

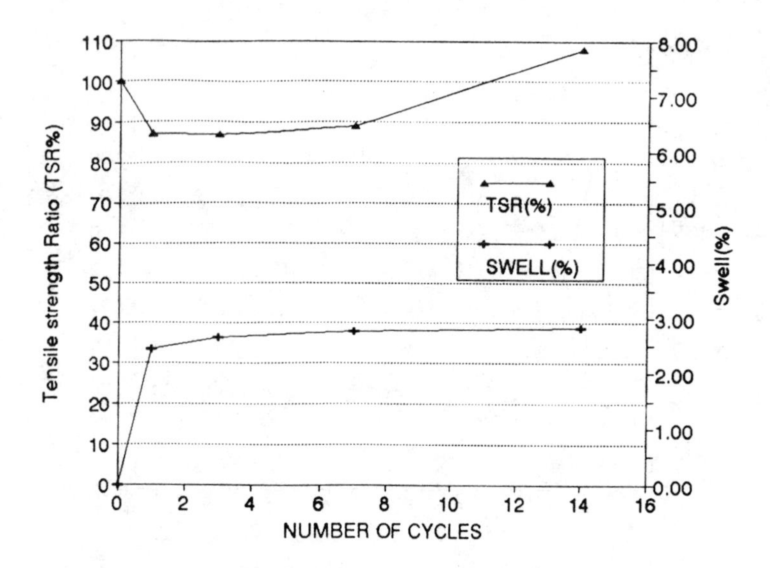

FIG. 4-- The extended Wet-dry Test Results for HMA with PCS #3.

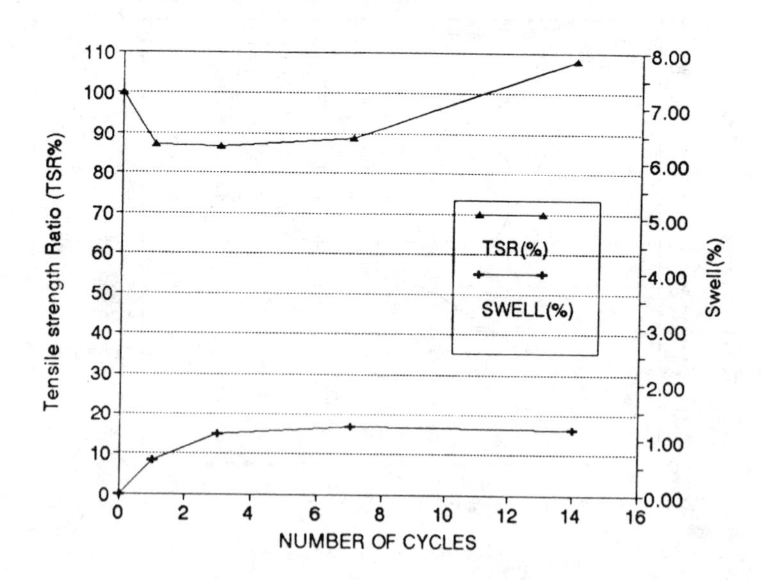

FIG. 5-- The extended Wet-dry Test Results for the Control Mix.

TABLE 6-- Durability of Laboratory HMA with PCSs.

HMA Mix	Wet/Dry TSR (%)	Freeze/Thaw TSR (%)
Control	91.7	82.3
HMA with Soil #1	98.0	89.0
HMA with Soil #2	89.3	100.0
HMA with Soil #3	87.2	93.9
HMA with Soil #4	83.8	87.0
HMA with Soil #5	93.4	98.4
HMA with Soil #6	100	100

TABLE 7-- Durability of Field HMA with PCSs.

Test #	Wet/Dry TSR (%)	Freeze/Thaw TSR (%)
1	93.0	80.0
2	99.0	76.0
3	N. D.	N. D.
4	100.0	96.0
5	81.0	79.0

Permeability of HMA with PCSs

For the permeability test, compacted HMA specimen was extruded from the Marshall compactor and was saturated by placing it under water inside a container subjected to a vacuum pressure. Then the saturated specimen was placed on top of a flexible-wall permeameter with 4 inch specimen diameter. Two porous stones were placed between the specimen and, top and bottom plates. The two plates were lightly coated with vacuum grease. A rubber membrane was fitted to the sample with two "O" rings. The membrane was thoroughly checked for possible leakages, by immersing it in water.

A cell pressure of 140 kPa was applied and the specimen was allowed to stabilize. Then inflow and outflow valves were closed and the cell pressure was increased to 344 kPa to apply a back- pressure of 204 kPa. Then the sample was permeated under a desired pressure difference (mainly 7 kPa). At this stage the cell pressure value was 344 kPa and the outlet and inlet pressures were 204 kPa and 211 kPa, respectively, if the pressure difference was 7 kPa. Twenty-four hours after the in-flow became equal to the out-flow, and when the hydraulic conductivity did not show a further reduction, the permeability test was stopped. The permeability test was conducted concurrently on three specimens of HMA made with same PCS. A bladder accumulator was connected between the permeameter and the pressure panel to separate the permeant from the distilled water used in the pressure panel. This procedure eliminated the contamination of the pressure panel. Each time the permeant was changed, the bladder accumulator was thoroughly washed and all the air bubbles in the bladder accumulator and the connecting tubes were flushed out.

The following measurements were made continuously while recording the duration of the test:

1. Outlet, inlet and chamber pressures in kPa.
2. Outlet, inlet and chamber volume changes in ml.
3. Temperature.

At the end of the permeability test, the specimen height was measured. The above measurements were used to calculate the change in hydraulic conductivity with time. The falling head and raising tail method was used to calculate the hydraulic conductivity values. Fig. 6 shows a typical permeability test result where the variation of hydraulic conductivity with time is shown. The hydraulic conductivity value reduced with time and reached a constant value after ten to fourteen days. These constant hydraulic conductivities of HMA specimens with six PCSs are shown in Table 8.

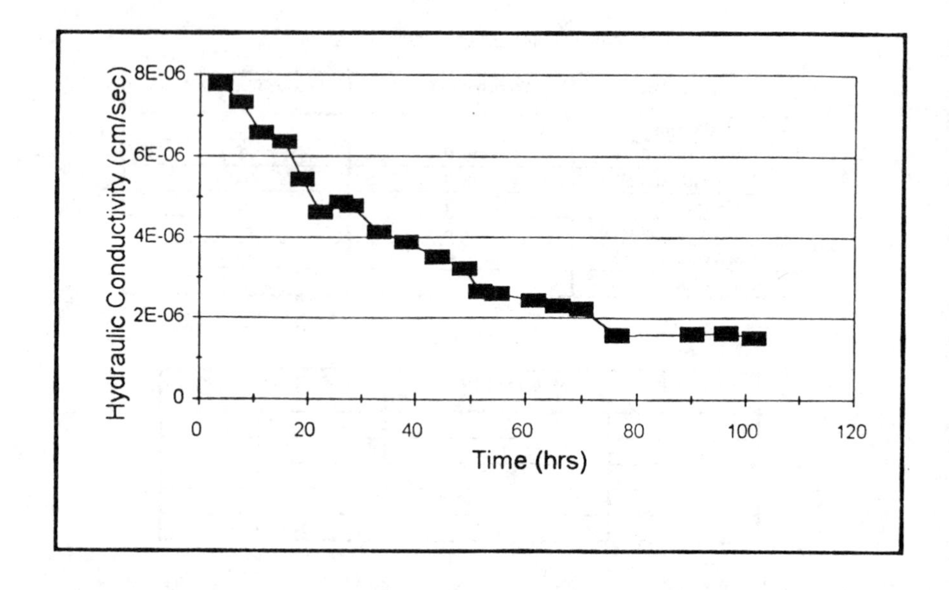

FIG 6-- A Typical Permeability Test Result of HMA with PCS, Variation of Hydraulic Conductivity with Time.

TABLE 8-- Hydraulic Conductivities of HMA with PCSs.

HMA Mix	Saturated Hydraulic Conductivity in cm/sec
Control	2.3E-07
HMA with Soil #1	3.3E-07
HMA with Soil #2	1.6E-07
HMA with Soil #3	1.6E-06
HMA with Soil #4	1.0E-06
HMA with Soil #5	8.3E-07
HMA with Soil #6	4.6E-07

The permeability tests were also performed on field compacted and hardened HMA concrete prepared with PCSs. Table 9 shows the hydraulic conductivity values for the HMA made with PCSs from the five field tests.

TABLE 9-- Hydraulic Conductivities of Field Compacted HMA with PCSs.

HMA Mix	Saturated Hydraulic Conductivity in cm/sec
Test #1	2.2E-07
Test #2	1.3E-04
Test #3	N/A
Test #4	1.0E-07
Test #5	5.0E-04

N/A: not applicable

DISCUSSION OF TEST RESULTS

Marshall Test

If an asphalt concrete meets all the state specifications and if it is a workable mix, then it is accepted as a paving material. Except for the control and PCS #3, as shown in Table 3, all the HMA with PCSs were acceptable as paving materials. The control mix and HMA with soil #3 had low flow values. However, a higher flow value can be selected for both from Marshall test results with higher asphalt contents, but it will result in lower stability values (lower than 8006 N). If a lower Marshall strength value (say 6670 N) is acceptable, even the control and HMA with soil # 3 are acceptable as paving materials. Based on the test results shown in Table 3 it may be stated that the HMA with PCSs produced better asphalt concrete when compared with the control. This may be due to the better blend obtained by adding natural soils. Very high Marshall stability values were obtained for HMA made with PCS #3 and #6. PCS #3 was a silty sand, and the well-graded size distribution produced a densely graded tight mix. This contributed to very high Marshall stability values. Though PCS #6 was a poorly graded sand with silt, it had a uniform size distribution of + D_{70} sizes. That uniform distribution distribution must have caused the higher Marshall stability values for this mix.

Based on the field test results (see Table 5) it may be concluded that test #1, which was an NJ I-3 mix, satisfied all the NJ specifications except the specification for flow. As stated earlier, a higher flow value can be obtained for higher asphalt content, but it may reduce the stability value. However, since the stability value was so high such a reduction should not be a concern. Other tests were based on NH specifications and were found to be satisfactory. It should also be noted that HMA produced from test #4 and #5 were used for paving and the paved sections of the low volume roads performed well after one year of service. The HMA produced from test # 3 was stored and later used as sub-base material. Since test #1 and #2 were field demonstrations, resulting HMA was recycled as RAP (recycled asphalt pavements).

Durability of HMA with PCSs

The extended durability test showed that one cycle of freeze & thaw and wet & dry was sufficient to evaluate the durability of HMA with PCSs. The percentage swell and the TSR computed can be used to evaluate the behavior of each mixture. Table 6 shows the TSR values of wet-dry and freeze-thaw tests for the control and for the HMA made with six soils. TSR values for HMA with PCSs were not significantly different from that of the control indicating that HMA with PCSs produced durable asphalt concrete. The freeze & thaw test results showed that the HMA with PCSs were sighly better than the HMA produced with virgin aggregates.

The durability of field test specimens was equally good as the lab compacted specimens indicating that even under field conditions the inclusion of PCSs in HMA does not reduce the durability of the hardened asphalt concrete.

Permeability of HMA with PCSs

Fig. 4 shows a permeability test result where the variation of hydraulic conductivity with time is plotted. It is clear from Fig. 6 that,as in the case of unsaturated soils, the hydraulic conductivity drops with the time associated with the increase in degree of saturation to yield a minimum hydraulic conductivity value that corresponded to 100% saturation. Table 8 shows the saturated hydraulic conductivity data for HMA with and without PCSs. Only one mix (with PC # 6) showed lower saturated hydraulic value than the control. However, the saturated hydraulic conductivity values of all the HMA mixes were less than 2.0E-06, acharacteristic value for silts and clays, and hence can be considered as acceptable. The field compacted samples showed similar results for test #1 and #4 as those of the lab tests . Test #2 and #5 had similar gradation and used for the base, hence it had higher hydraulic conductivity values. Since test #3 was used as filling product, it was not possible to obtain specimens for the permeability test.

SUMMARY AND CONCLUSIONS

Traditionally five to ten percent of industrial waste products are added to hot mix asphalt concrete without hurting the asphalt concrete matrix. The soils contaminated from leaking USTs, usually treated as solid waste, have been added to HMA since 1985. The quantity of such soils are projected to increase substantially over the next few years. In this paper the feasibility of using petroleum contaminated soils (PCSs) in the production of asphalt concrete is discussed. When PCSs are used in asphalt production three beneficial actions occur: incineration, dilution and solidification.

Laboratory and field studies were performed to evaluate the acceptability of PCSs in HMA. The impact with respect to strength of asphalt concrete due to the addition of PCS to HMA was evaluated by performing Marshall stability test. The Marshall stability tests for lab and field compacted specimens showed that the addition of PCSs produced acceptable asphalt concretes which were within the state specifications. The freeze & thaw and wet & dry tests were performed to test the durability of HMA made with PCSs. The durability tests showed that the HMA with PCSs were as durable as the control. In addition to the above the hydraulic conductivities of HMA with PCSs were determined. The hydraulic conductivity of HMA with all PCSs produced asphalt concrete with hydraulic conductivities less than 1.0E-06. Based on the above test data it can be concluded that HMA made with PCS is suitable as a paving material based on current specifications.

ACKNOWLEDGMENTS

The authors wish to acknowledge 1) the State of New Jersey Department of Environmental Protection and Energy for funding the research described in this paper, and 2) Mr. Mark Charbonneau of Continental Paving Co., NH and Mr. Mike Manno of Newark Asphalt Co., NJ for their support in performing the research described in this paper.

REFERENCES

[1] Fairweather, V., "US Trackless Leaking Tanks" Civil Engineering ASCE , 1990

[2] Czarnecki, R., "Making Use of Contaminated Soils", Civil Engineering, ASCE, Dec. 1988, pp. 72-74.

[3] Czarnecki, R., "Hot Mix Asphalt Technology and Cleaning of Contaminated Soils", Petroleum Contaminated Soils, Vol. II, Editors, P. Kostecki and E. J. Calabrese, Lewis Publishers, Inc., 1989.

[4] USDOT, Selected Highway Statistics and Charts, Federal Highway Administration, US Department of Transportation, 1988.

[5] Asphalt Institute, The Asphalt Handbook, Asphalt Institute, MS-4, 1989.

[6] Meegoda, N. J., Huang, D. R. DuBose, B. H., and Chen, Y., "Use of Petroleum Contaminated Soils in Construction Material Production", Interim Report to the New Jersey Department of Environmental Protection, Division of Science and Research, Submitted by the New Jersey Institute of Technology, Newark, NJ, August 1991, pp. 1-80.

[7] Meegoda, N. J., Huang, D. R., DuBose, B., and Mueller, R. T., "Use of Petroleum Contaminated Soils in Asphalt Concrete", Hydrocarbon Contaminated Soils Volume II, Chapter 31, pp 529-548. (P. T. Kostecki, E. J. Calabrese & M. Bonazountas, Editors), Lewis Publisher, 1992.

Polyethylene Waste

Zhi-zhong Liang[1], Raymond T. Woodhams[1], Zhen N. Wang[1] and Bruce F. Harbinson[2]

UTILIZATION OF RECYCLED POLYETHYLENE IN THE PREPARATION OF STABILIZED, HIGH PERFORMANCE MODIFIED ASPHALT BINDERS

REFERENCE: Liang, Z., Woodhams, R. T., Wang, Z. N., and Harbinson, B. F., **"Utilization of Recycled Polyethylene in the Preparation of Stabilized, High Performance Modified Asphalt Binders,"** Use of Waste Materials in Hot-Mix Asphalt, ASTM STP 1193, H. Fred Waller, Ed., American Society for Testing and Materials, Philadelphia, 1993.

ABSTRACT: This paper proposes a unique, and proprietary, combination of a polyethylene and an elastomer for the modification of paving asphalts in order to minimize low temperature cracking and reduce rutting at elevated seasonal temperatures under heavy loads. A key objective of the research was to demonstrate the effectiveness of recycled polyethylene (PE) as a material for the modification of paving asphalts. The results indicate that the technology has the potential to create a sustainable downstream market for the safe diversion of large amounts of low grade post consumer polyethylene. Minor contaminants and mixed coloration of these waste materials pose difficulties for many potential applications but show no significant adverse effects on the quality of the modified asphalts. In addition to creating an important outlet for waste plastics, substantial material cost savings are possible over virgin polyethylene.

Asphalts modified by this process contain polyethylene particles which remain emulsified by an elastomeric steric layer. The liquid emulsion remains permanently stable at elevated temperatures and is not adversely affected by repeated heating/cooling cycles.

The low temperature fracture toughness of several asphalts modified with different grades of polyethylene was improved at temperatures near -20°C when compared to that of unmodified asphalt and another polymer modified asphalt. These same compositions also exhibited greater stiffness and elasticity at elevated temperatures near 60°C. Mix design and performance for selected stabilized PE modified binders were also enhanced when compared with unmodified asphalt.

The incentive for this research is the prospect of paved roads having substantially longer service lifetimes using environmentally neutral waste polyethylenes as modifiers.

KEYWORDS: asphalt, recycled polyethylene, elastomer, steric stabilization, direct compression test, low-temperature fracture, shear rheometer, dynamic rheological property

---

[1]Research Associate, Professor, and Research Assistant respectively, Department of Chemical Engineering and Applied Chemistry, University of Toronto, Toronto, Canada M5S 1A4

[2]President, Polyphalt Inc., 200 College St., Toronto, Canada M5S 1A4

It is well known that the addition of certain polymers to asphalt binders can improve the performance of paved roads [1-6]. Polyethylene has been found to be one of the most effective polymer additives by virtue of its low glass transition temperature and crystalline structure [7-12]. It is also the least expensive due to an abundant supply of waste or recycled polyethylene. However simple mixtures of molten polyethylene and asphalt were found to be unstable during storage and transport and would rapidly coalesce and separate into layers unless continuously stirred. This disadvantage has limited the commercial use of polyethylene modified asphalt binders in pavements. Therefore, an objective of this research was to develop a polyethylene-modified asphalt which could remain permanently stable at elevated temperatures and would not phase separate or change its viscosity during long term storage [13].

It has been found [14] that molten polyethylenes may be conveniently emulsified in asphalt using minor quantities of an elastomeric emulsifying agent, commonly referred to as a steric stabilizer. The liquid emulsion remains permanently stable at elevated temperatures and is not adversely affected by repeated heating and cooling cycles. The process can employ a wide variety of virgin or recycled polyethylenes in most asphalts. Since the asphalt formulations may employ recycled polyethylene waste, without adversely affecting quality, substantial material cost savings are possible. The incentive for this research is the prospect of paved road having substantially longer service lifetimes using environmentally neutral waste polyethylenes as modifiers.

It has been demonstrated in paving trials that asphalt concretes which employ polyethylene modified binders are more resistant to rutting during elevated seasonal temperatures [8]. The low temperature fracture toughness of several asphalts modified with different types of polyethylene was evaluated at temperatures near -20°C and compared to the corresponding unmodified asphalts. The viscoelastic properties were measured at 60°C with a shear rheometer in order to predict elevated temperature performance. Selected samples of these new stabilized PE modified binders were also compared with unmodified asphalts for mix design and low temperature performance characteristics.

EXPERIMENTAL

Materials

Ishtmasmaya (Ishtm) and Lloydminister (Lloydm) asphalt binders were selected for performance evaluations at low temperatures. The Ishtm binder (85/100) is a residual Mexican bitumen obtained from Petro-Canada. For mix design tests and high temperature evaluations of binder performance, Lloydm asphalts (85/100 and 150/200 penetration) were also selected. The physical properties of these asphalts are presented in Table 1.

Virgin and several types of virgin and recycled polyethylene materials were examined during the course of this investigation. Detailed information on the individual polyethylenes used are shown in Table 2. The stabilized PE modified asphalts were prepared in a one-litre heated reactor vessel using a high shear mixer. The various formulations are summarized in Table 3. A commercial rubber modified binder (designated CRMB) has been included for comparison purposes only. Since the choice of asphalt can influence the mechanical properties of the resulting modified binder, it is important to evaluate performance variations primarily with respect to the parent asphalt. Typical

TABLE 1--Properties of base asphalts.

Property	Test Method	Asphalt Type		
		Lloydm-1	Lloydm-2	Ishtm
Penetration grade		85/100	150/200	85/100
Viscosity, cst at 135°C	ASTM D2070	383	--	320
Density at 15°C	Asphalt Institute[a]	1.026	1.025	1.029

[a]Asphalt Technology and Construction Practices, ES-1, B50, 2nd Ed., Jan. 1983.

TABLE 2--Polyethylene additives.

Polymer Type	Description
Virgin PE	Sclair 8409 linear low density polyethylene (LLDPE) from Dupont, Melt Index = 12 g/10 min, (ASTM D1238) low-temperature brittleness point <-70°C, (ASTM D746)
R-PE-1	Recycled LLDPE waste (Colortech Ltd.). Melt Index about 0.8. Solids content is 47% by weight (by wt) including titanium dioxide, calcium carbonate and carbon black.
R-PE-2	Post consumer high density polyethylene mixed colour flake (Resources Plastics Corp.); Melt Index about 0.5; contains less than 5% polypropylene (PP) and less than 0.5% polyethylene terephthalate (PET) and polystyrene combined.
R-PE-3	Recycled commingled polyolefin pellets from diaper waste. Contains approximately: LLDPE--60% by wt; PP--40% by wt and; TiO_2--1 to 2% by wt.

TABLE 3--Polymer modified asphalts.

Code	Polymer Type	PE (wt. %)	Stabilizer (wt. %)	Base Asphalt
PMB-1	Virgin PE	3	1	Lloydm-1
PMB-2	Virgin PE	4	2	Lloydm-1
PMB-3	Virgin PE	4	1	Lloydm-1
PMB-4	Virgin PE	4	2	Lloydm-2
PMB-5	Virgin PE	3	1	Ishtm
R-PMB-1	R-PE-1	3	1	Ishtm
R-PMB-2	R-PE-2	3	1	Ishtm
R-PMB-3	R-PE-3	3	1	Ishtm
CRMB[a]	SB rubber	--	--	unknown

[a] a commercial asphalt binder modified with styrene-butadiene rubber.

TABLE 4--Typical properties of stabilized PE modified asphalt binders.

Property	ASTM Test Ref.	PMB-1	PMB-2	Control[a]	CRMB
Penetration at 25°C	D5	64	63	82	71
Viscosity, mm^2/s					
at 60°C	D2171	3 446	4 824	1 878	5 060
at 135°C	D2070	1 154	1 493	383	742
After TFOT[b]	D1754				
Penetration at 25°C		43	43	57	49
Penetration % Retained		67	68	69	69

[a] Lloyd 85/100.
[b] thin film oven test.

properties of two stabilized PE modified asphalt binders are shown in Table 4. These samples were selected as being representative of the minimum (PMB-1) and maximum (PMB-2) levels of polymer modification.

Evaluation

Morphological analyses of the polymer modified asphalts were conducted using a microscope equipped with a hot-stage. A molten drop of binder was placed between microscope slides and pressed into a thin film which could be viewed by transmitted light. The individual polymer particles suspended in the asphalt medium could be seen in motion and photographed at intervals.

The low temperature viscous flow behaviour of the asphalt binders was evaluated according to the ASTM 695 compression test. The cylindrical asphalt specimens (height = 0.64 cm, diam. = 1.27 cm) were compressed in an environmental chamber attached to an Instron Tester. The crosshead speed was 1.27 mm/min for all tests. Each result was reported as the average of at least 5 repeat specimens. The ASTM compression test was found to be more reproducible than a corresponding flexural test which is a complex interaction of shear, compression and tension. The compression test was also more sensitive to the transition temperature from plastic to brittle behaviour at various crosshead speeds.

Performance of the modified binders at elevated temperatures near the softening point was determined with a dynamic shear rheometer (Bohlin) using a cone and plate geometry. The gap height was set at about 0.8 mm. Selection of platen diameter is related to the stiffness of the sample (ie. AC grade); a 12.5 mm platen diameter was found satisfactory for the testing of these materials at 60°C. Measurements were confined to the linear viscoelastic region by repeating frequency sweeps at two different strains in which it was assumed $G''(\omega)$ does not equal $f(\gamma)$ (for detailed information on asphalt dynamic shear testing refer to [15]). The mix designs were carried out using the HL-1 mix design (Ontario Ministry of Transportation Contract 90-07).

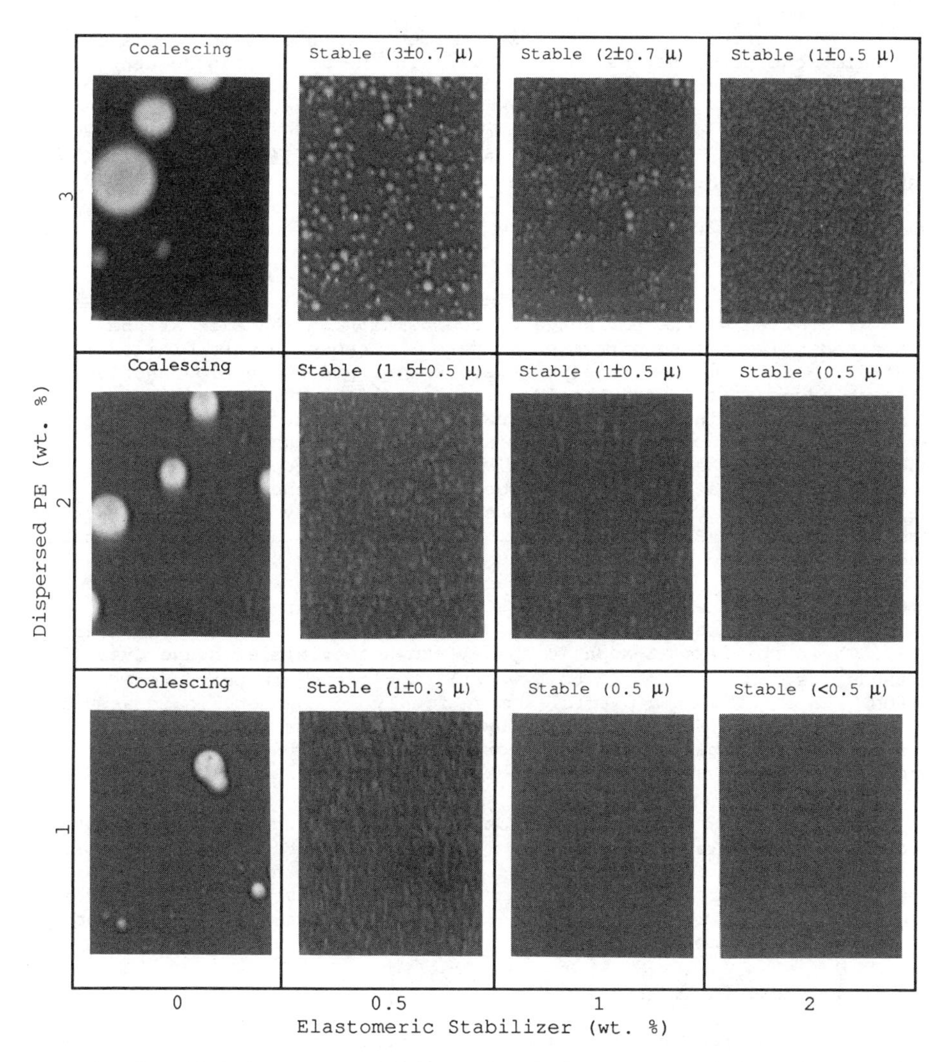

FIG 1--Morphology of sterically stabilized PE modified binders: status (particle size).

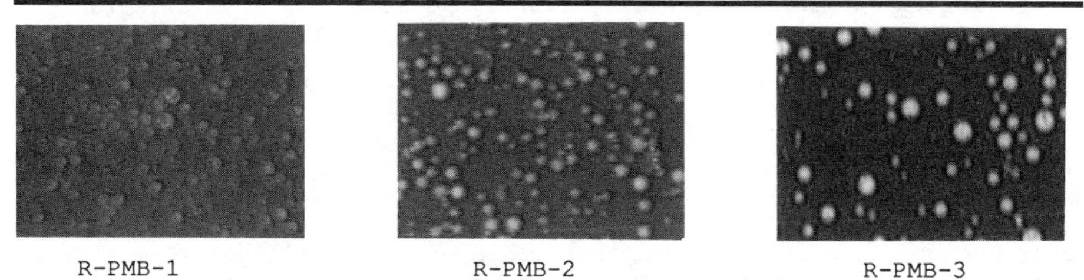

FIG 2--Comparative morphology of stabilized binders using different recycled PE.

The modified asphalt mix was compared with the standard mix for Marshall stability, Marshall flow, voids in the mineral aggregate (VMA) and air voids according to ASTM D1559. The asphalt mix performance at low temperature (-18°C) was evaluated at the University of Waterloo using a special constant rate extension tensile test. This test simulates stresses imposed by thermal contraction [16]. Each value was determined from the average of four repeat tests.

RESULTS AND DISCUSSION

Morphology

Fig. 1 shows the influence of composition on the size of the dispersed polyethylene particles compared to those in unstabilized controls. In the molten PE-asphalt system (virgin PE in Lloyd 85/100) without added stabilizer, the PE particles (light colour) can be observed to immediately begin coalescing. Note that the elastomeric stabilizer is soluble in the bituminous continuous phase (dark background) and, hence, is not visible. The continued coalescence of the dispersed phase leads to a large reduction in the number of particles and eventual complete phase separation. By carefully adjusting the proportion of PE and elastomeric stabilizer in the modified asphalt, the average particle size can be controlled to less than one micron in diameter. Particle size and interparticle distances are known to have important influences on the yield stress and fracture toughness of composites in which an elastomer has been dispersed in a brittle matrix [17]. The stabilized PE-asphalt system shows no visible change over extended periods of time (2 to 4 weeks) at elevated temperatures (160°C). The stabilized particles move freely in the hot liquid asphalt and are prevented from coalescing by the elastomeric steric barrier surrounding each particle. The theory of steric stabilization in asphalt-polymer systems has been outlined in a previous paper [13].

Fig. 2 shows PE modified binders prepared using three different recycled polyethylenes (Table 2) in a Lloyd 85/100 base asphalt. Apart from minor morphological variations resulting from the level of impurities, fillers and other polymers (R-PMB-3) these modified binders showed no appreciable difference in morphology or stability from binders prepared using virgin PE.

Low Temperature Performance

It is instructive to observe the complete force-displacement response of an asphalt specimen when it is subjected to a compression stress at low temperatures. Fig. 3 illustrates typical displacement curves for two unmodified asphalts (85/100 penetration) from different sources (Lloydm and Ishtm) and the corresponding stabilized PE modified asphalts. For a particular strain rate (related to the cross-head speed), these two asphalts showed a distinct difference in fracture mode at -15°C. The unmodified Lloydm asphalt is brittle at this temperature and fails in a catastrophic mode. The Ishtm asphalt exhibited a ductile deformation at -15°C, therefore, a lower temperature (-20°C) was selected to ensure brittle fracture for the control sample so that any significant increase in ductility attributable to polymer additives could be readily distinguished. Comparison of the stabilized PE-modified asphalts with the controls in Fig. 2 shows a pronounced increase in both yield strength and failure strain under these severe conditions. Figs. 4 and 5 reveal the low temperature performance of several stabilized PE

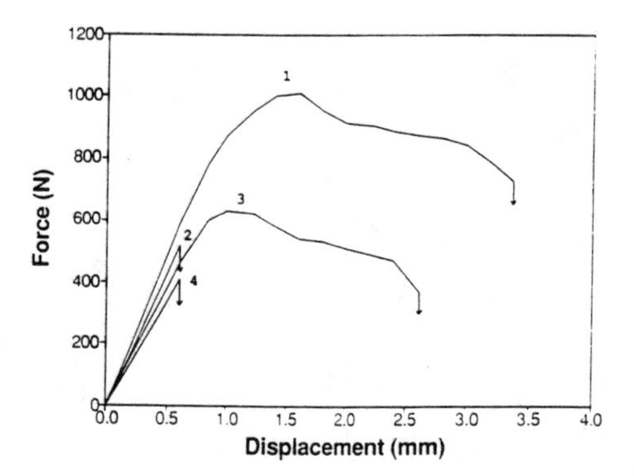

FIG. 3--Typical force-displacement curves of PE modified asphalts.
1.--Modified Ishtm at -20°C (PMB-5) 2.--Ishtm control
3.--Modified Lloyd at -15°C (PBM-1) 4.--Lloyd control

modified asphalts compared to a commercial styrene/butadiene rubber modified binder. The stabilized polyethylene modified binders all exhibited high yield strength and failure strain relative to the asphalt control, indicating improved fracture toughness at -20°C. It is also significant that the recycled PE materials were virtually equivalent in effectiveness when compared to virgin polyethylene (PMB-5) for enhancing the low temperature performance of asphalt. PE modified asphalts appear to compare favorably with the styrene/butadiene type modifier.

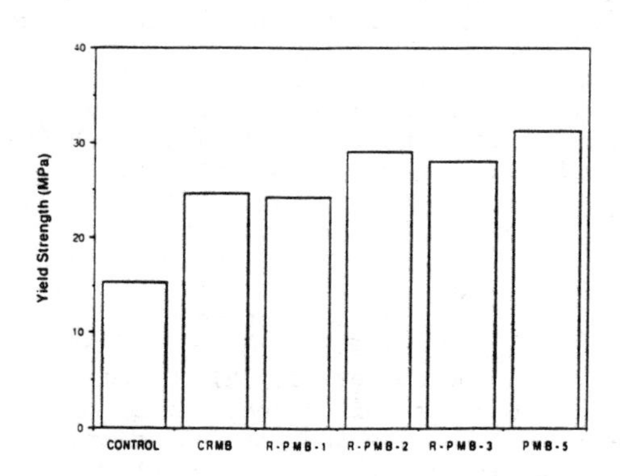

FIG. 4--Yield strength of modified asphalt binders at -20°C.

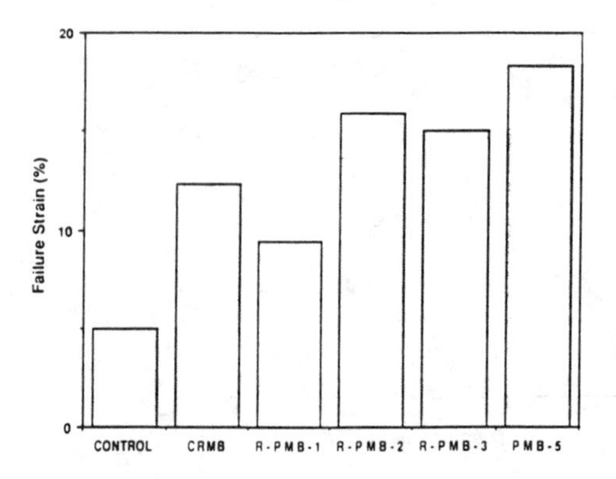

FIG. 5--Failure strain of modified asphalt binders at -20°C.

A comparison among the three different types of recycled PE (Figs. 4 and 5) indicated that some differences in composition of waste materials were reflected in the properties of asphalts modified with them respectively. The recycled LLDPE (R-PMB-1) contained pigments and fillers such as titanium dioxide, calcium carbonate and carbon black. These inorganic materials do not appear to have a major effect on the yield strength properties, however failure strain is below that of the other PE modified binders. The polyethylene in sample designated R-PMB-2 was derived from post-consumer polyethylene waste. Residual contamination does not appear to have influenced the results. While research to date has been conducted primarily with polyethylene material, the recycled diaper waste (R-PMB-3) appears equally effective despite the presence of approximately 40% (by wt) polypropylene. It is possible to emulsify most forms of polyethylene with the exception of crosslinked polymers.

The improved low temperature yield performance of the stabilized PE modified binders is thought to be a result of synergistic benefits between the PE core and the elastomeric steric layer. Polyethylene has a much lower glass transition temperature than the asphalt matrix in which it is dispersed as does the elastomeric stabilizer. Moreover, the stabilizer behaves like an emulsifier, spontaneously creating PE particles which are much smaller than can be achieved using high shear mixing alone. It is well recognized from the polymer industry that the ability to control the particle size of elastomer dispersed in a brittle matrix at the micron level is particularly important for imparting increased toughness at low temperatures [17,18]. Similarly, the elastomeric stabilizer, in the case of these modified binders, facilities control over particle size and particle dispersion characteristics, such as inter particle distance, which in turn may enhance performance at low temperatures.

High Temperature Performance

Thermal stability of the modified asphalt binders was examined with a Bohlin dynamic shear rheometer at a temperature of 60°C; typically the maximum service temperature of asphalt roads.

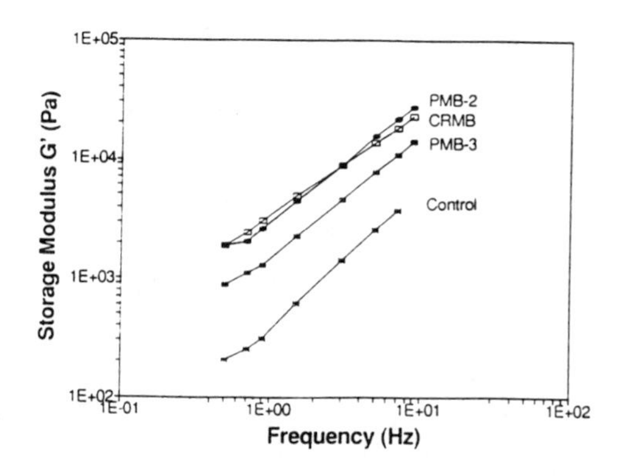

FIG. 6--Storage modulus versus frequency at 60°C.

Fig. 6 compares the elastic storage modulus (G') of two stabilized PE modified binders (PBM-2 and PBM-3) with the parent asphalt control and a commercial SB rubber modified asphalt, across a broad range of frequencies approximating normal traffic conditions [15,19,20]. Both stabilized PE modified binders exhibited a greater storage modulus (a measure of elasticity) compared to the asphalt control. Sample PBM-2 exhibits a storage modulus similar to that of the commercial SB rubber modified binder.

The complex dynamic viscosity (η^*) of the same four binders is shown in Fig. 7 as a function of frequency at 60°C. The dynamic viscosity provides an indication of flow behaviour. It is evident from Fig. 7 that stabilized PE asphalts have a greater viscosity than the corresponding unmodified control at this temperature, which should result in enhanced ability to resist permanent deformation. The SB rubber modified asphalt appears more sensitive to frequency than the polyethylene systems.

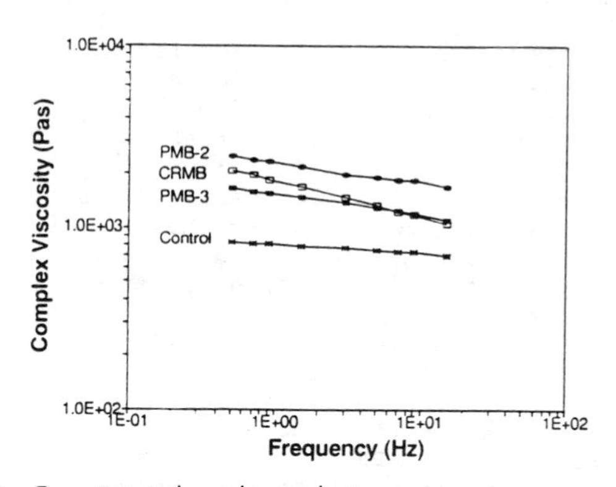

FIG. 7-- Dynamic viscosity versus frequency at 60°C.

According to the latest version [19] of the binder performance specifications developed by the Strategic Highway Research Program (SHRP), these viscoelastic parameters can be correlated with high temperature binder performance with the expression $G^*/\sin\delta$ (G^*: complex

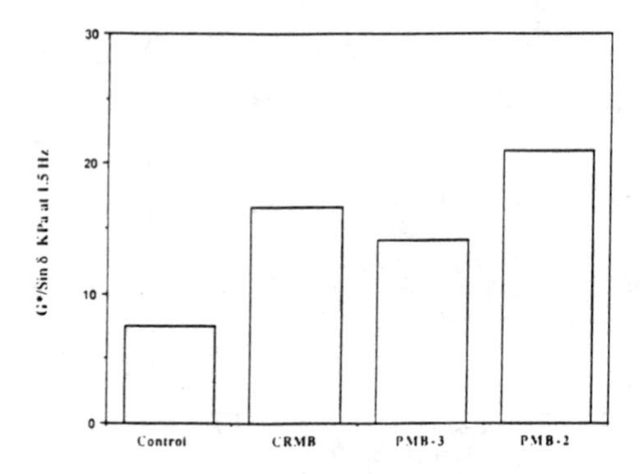

FIG. 8--Viscoelasticity of asphalt binders at 60°C before TFOT.

modulus and δ: phase angle). The values of G*/sinδ for various asphalts are related to tenderness and rutting before and after aging. This function is presented in Fig. 7 for the same materials at 1.5 Hz and 60°C. The numerical results predict increased dimensional stability and greater resistance to rutting when compared to the unmodified asphalt control. The styrene/butadiene rubber modified asphalt has an intermediate value.

The dynamic rheological properties of sample PMB-2 consistently exceed those of PMB-3 in Figs. 5, 6 and 7. This is likely attributable to an increased level of elastomeric stabilizer which, in addition to its own beneficial properties, has a morphological effect, helping to create a reduction in PE particle size and inter particle distance as shown in Fig. 1. Generally, these dynamic rheological data predict greater dimensional stability for stabilized PE modified asphalts at elevated service conditions when compared to unmodified asphalts.

Mix Design and Performance

The Marshall test results for a mix prepared with stabilized PE asphalt binder (PMB-4 with 150/200 penetration asphalt) are summarized in Fig. 9 and Table 5. Comparison with a standard Petro-Canada asphalt mix (85/100 penetration) provided assurance that the stabilized PE modified binder is similar in processability and behaviour. This mix provides increased Marshall Stability at the same Marshall Flow value. The remaining air voids after compaction and the voids in the mineral aggregate (VMA) are similar to the HL-1 asphalt concrete control, indicative of a similar degree of compaction for these mixes.

The low temperature performances of the mixes constructed with the modified binders (PMB-1 and PMB-2) are shown in Fig. 10. The PE modified asphalt mixes exhibited high mean tensile strengths (723 and 727 psi), which were 25% greater than the asphalt control. These increased tensile strengths, combined with a 40 percent increase in failure strains demonstrate that these stabilized PE modified mixes should have greater toughness and durability in service. These results on the asphalt concrete mixes are consistent with the observed behaviour of the corresponding binders as described earlier in this paper.

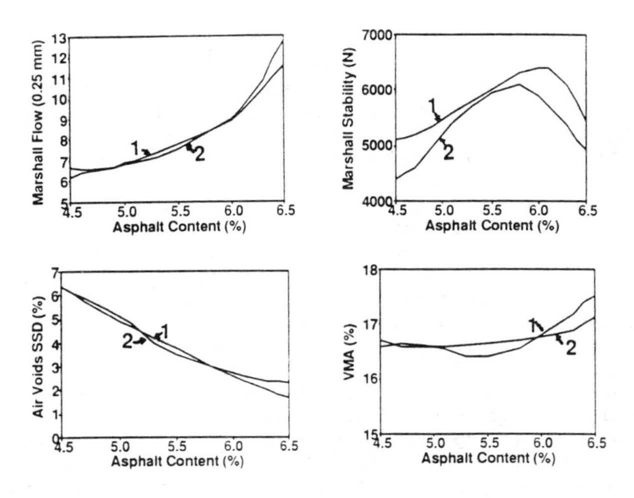

FIG. 9--Marshall mix design results (1--Mix constructed with the PMB-4 asphalt and 2--Mix constructed with a 85/100 penetration asphalt).

TABLE 5--Marshall test results.

Property	Unmodified 85/100 base	Modified (PMB-4) 150/200 base
Air voids (%)	4.9	5.0
VMA (%)	16.6	16.6
Marshall Flow (0.25 mm)	6.8	6.7
Marshall Stability (N)	5 186	5 477
Asphalt Content (%)	5.0	5.0

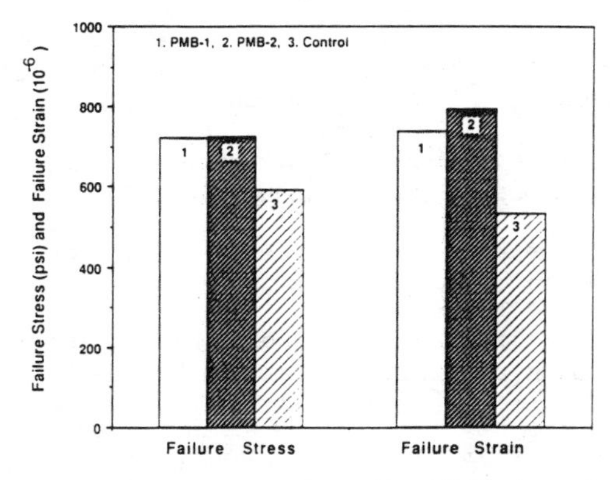

FIG. 10--Low temperature properties of modified asphalt mixes at -18°C.

CONCLUSIONS

Laboratory evaluations on binder and mix samples confirm the well known effectiveness of polyethylene as an additive for increasing the resistance of asphalt towards low temperature cracking, and also reduced rutting deformation at elevated temperatures. Elastomeric steric stabilizers were employed to permanently emulsify the polyethylene in asphalt and prevent phase separation during heated storage or transport. The elastomer based steric stabilizer permits the size of the emulsified polyethylene particles to be accurately controlled to micron dimensions, thereby providing precise control of low temperature mechanical properties while also contributing to dimensional stability at elevated service temperatures. The process is applicable to most grades of polyethylene, including mixed PE, and most asphalt types. Dynamic viscoelastic properties of the modified binders, according to current SHRP specifications, predict an excellent balance of processability, strength, toughness and resistance to deformation. Moreover, the technology presents an important opportunity for adding value to large quantities of low grade post-consumer polyethylene waste materials. Further research is now being directed at paving trials in order to confirm the expected increase in performance and service life of roads exposed to extreme climates and heavy traffic.

ACKNOWLEDGEMENTS

This research was partially sponsored by the Ontario Ministry of Transportation, The Natural Sciences and Engineering Research Council of Canada (Strategic Grant No. 45408) and the Ontario Centre for Materials Research. The low temperature test results (asphalt concrete mix) were provided by Dr.Ralph Haas of the University of Waterloo and the low temperature tests for the binder by Mr. Zhen Wang, Research Assistant. Appreciation is also expressed to Dr. John Vlachopoulos and McMaster University for use of rheometric equipment, to Dr. Simon Hesp for advice on sample preparation (PMB-4) and to the following organizations for their contribution of materials: Resource Plastics Corp., Colortech Ltd. and Petro-Canada.

REFERENCES

[1] Woodhams, R. T., "Methods of Increasing the Fracture Toughness of Asphalt Concrete," in Transportation Research Board Record 843, National Academy of Sciences, 1982, pp. 21-26.

[2] Button, J. W., Little, D. N., Kim, Y. and Ahmed, J., "Mechanistic Evaluation of Selected Asphalt Additives," in Proc. of the Association of Asphalt Paving Technologies, Vol. 56, 1987, pp. 62-90.

[3] Lee, D. Y. and Demirel, T., "Beneficial Effects of Selected Additives on Asphalt Cement Mixes," Final Report, Project HR-278, Iowa Department of Transportation, 1987.

[4] King, G. N. and King, H. W., "Polymer Modified Asphalt - An Overview," in Solutions for Pavement Rehabilitation Problems, LaHue, S. P. Ed., New York ASCE , 1982, pp. 240-254.

[5] Zanzotto, L., Foley, D., Watson, R. D. and Juergens, C., "On Some Practical Aspects of Using Polymer Asphalts in Hot Mixes," Canadian Technical Asphalt Association Proceedings, Vol. 34, 1989, pp. 20-40.

[6] Kraus, G., "Modification of Asphalt by Block Copolymers of Butadiene and Styrene," Rubber Chem. Tech., Vol. 55, No. 5, 1982, pp. 1389-1402.

[7] Jew, P. and Woodhams, R. T., "Polyethylene Modified Bitumens for Paving Applications," Proc. of the Association of Asphalt Paving Technologies, Vol. 55, 1982, pp. 541-63.

[8] Denning, J. H. and Carswell, J., "Assessment of Novophalt as a Binder for Rolled Asphalt Wearing Course," TRRL Report 1101, Transport and Road Research Laboratory Crowthome (England), 1983.

[9] Haberl, P., "Polyethylene-Bitumen Compositions for Pavement," Ger. Offen. DE 2146915, March 1972.

[10] Runa, A., Sekera, M. and Masarykova, M., "Use of Polyethylene in the Manufacture of Modified Paving Asphalts and Experience With Their Use in Czechoslovakia," Ropa Uhilie, Vol. 29, No. 9, 1987, pp. 529-44.

[11] Moran, L. E., "Polyethylene Modified Asphalts," U.S. Pat., US 4868233, Sep. 19, 1989.

[12] Woodhams, R. T., "Bitumen-Polyolefin Compositions," PCT Int Appl. WO 87/5313 A1, Sep. 1987.

[13] Hesp, S. and Woodhams, R. T., "Stabilization Mechanisms in Polyolefin-Asphalt Emulsions," Polymer Modified Asphalt Binders, ASTM STP 1108, K. R. Wardlaw and S. Shuler, Eds., American Society for Testing and Material, Philadelphia, 1991.

[14] Hesp, S., Liang, Z. and Woodhams, R. T., "Bitumen-Polymer Stabilizer, Stabilized Bitumen-Polymer Compositions and Methods for the Preparation Thereof," US Patent application, Aug. 1991.

[15] Collins, J. H., Bouldin, M. G., Gelles, R. and Berker, A., "Improved Performance of Paving Asphalts by Polymer Modification," Proc. of the Association of Asphalt Paving Technologies, Vol. 60, 1991, pp.43-79.

[16] Kallas, Bernard F., "Low-Temperature Mechanical Properties of Asphalt Concrete," Research Report 82-3, The Asphalt Institute, Sep. 1982.

[17] Wu, S., "A Generalized Criterion for Rubber Toughening: The Critical Matrix Ligament Thickness," Polym. Prepr., American Chemistry Society, Vol.28, No. 2, 1982, pp. 179-80.

[18] Riew, C. K., Rowe, E. H. and Siebert, A. R., Toughness and Brittleness of Plastics, Chapter 27, 1976.

[19] Goodrich, J. L., "Asphaltic Binder Rheology, Asphalt Concrete Rheology and Asphalt Concrete Mix Properties", Proc. of the Association of Asphalt Paving Technologies, Vol. 60, 1991, pp. 80-120.

[20] Youtcheff, J., "Update on the SHRP Binder Specification - Draft 7G," Focus, Strategic Highway Research Program, Aug., 1992, pp. 1-2.

Dallas N. Little[1]

ENHANCEMENT OF ASPHALT CONCRETE MIXTURES TO MEET STRUCTURAL REQUIREMENTS THROUGH THE ADDITION OF RECYCLED POLYETHYLENE

REFERENCE: Little, D. N., "**Enhancement of Asphalt Concrete Mixtures to Meet Structural Requirements through the Addition of Recycled Polyethylene,**" Use of Waste Materials in Hot-Mix Asphalt, ASTM STP 1193, H. Fred Waller, Ed., American Society for Testing and Materials, Philadelphia, 1993.

ABSTRACT: Low density polyethylene is a major part of the waste plastic stream. Its use as a recycled additive in hot mix asphalt concrete pavement extends back to the mid-1970's in Europe. Recent and extensive testing has demonstrated that waste or recycled polyethylene is as effective in modifying the hot mix asphalt concrete as is virgin polyethylene as long as the recycled polymer is free of metals and paper and the percentage of additional polymer does not exceed 17 percent. The resistance to deformation of hot mix asphalt concrete modified with approximately 5 percent low density polyethylene is significantly better than that of the unmodified mix. Results of uniaxial creep testing parameters, including creep modulus, slope of the steady state portion of the creep curve and time to tertiary creep, on a large number of mixtures are used to evaluate the influence of recycled polyethylene modification. Moderate and low temperature fatigue properties are also evaluated based on controlled stress and controlled deformation fracture fatigue testing.

KEYWORDS: hot mix, polymer, polyethylene, modification, additives, creep test, permanent deformation, fatigue, rutting, pavement performance

RECYCLED POLYOLEFINS IN ASPHALT CONCRETE

The Need to Recycle Polyolefins

There is little debate that a need exists to recycle plastics which now comprise approximately 8.3 percent of the average composition of U. S. Trash (1). Indeed plastics recycling is on the rise in the U. S. In 1990 the United States recycled approximately 400,000 tons (363 million kg) of its 16.2 million tons (14.6 trillion kg) of municipal plastic waste - twice what was recycled in 1988. However, this is still a paultry 2.5 percent.

[1] Kelleher Professor of Civil Engineering, Texas A&M University

The Environmental Protection Agency's 1992 update of its biennial Characterization of Municipal Solid Waste in the United States shows that although plastics recycling is on the increase, the 1990 waste stream of plastics is over 40 times greater than the 400,000 tons (363 million kg) of plastics wasted in 1960 (2). The increase in the amount of plastics recycling is, of course, positive. However, the volume actually recycled compared to the waste stream is minuscule.

Polyolefin-Modified Asphalt Binders

In 1982, the Richard Felsinger Company of Vienna, Austria, began incorporating recycled resins in commercial asphalt paving applications in Europe. The process had been developed by the Felsinger Group in 1976. To date the Felsinger Company claims to have placed about 20 million pounds (9.07 million kg) of plastic waste to use in binder systems on a world-wide basis. The current plan is to use 92 percent recycled resin and 8 percent virgin material.

Typically 4 to 6 percent of either virgin or recycled polyolefins are used to modify asphalt binders and to improve temperature susceptibility. If 5 weight percent of recycled plastic were included in all asphalt paving mixtures produced in the United States, about 2.5 billion pounds (1.13 trillion kg) of recycled plastics could be used each year.

The Use of Polyolefins and Other Polymers as Additives in Paving Grade Asphalt

Polyethylene has been used almost exclusively as the polyolefin of choice as an asphalt additive. Both virgin and recycled polyethylene has been found to yield very good results. In fact, in 1989 a study was conducted by the Felsinger Group in Austria to evaluate the effects of minor proportions of other plastic materials on the performance of polyethylene modified asphalt binder. The study concluded that recycled low density polyethylene (LDPE) can be used to produce a premium binder - equally as good as one produced with virgin LDPE. The study further concluded that other polymeric substances added at concentrations of up to 17 percent of the polyethylene do not significantly influence the blending process or the properties of the modified binder in the hot mix[2]. In this study the polymer substances included: polypropyle, PET, BVP, polystyrene and PVC. The study also demonstrated that high density polyethylene when incorporated into asphalt cement causes an excessive increase in the Ring and Ball Softening Point and an unacceptable decrease in the penetration. If this binder is aged, the penetration value would be well below that which is accepted in the U. S. The aged binder would in all likelihood be very brittle and thus induce premature cracking in pavements. Figure 1 is a summary of the effects of other polymer substances in addition to the LDPE on the binder properties of Softening Point, Frass Break Point and Penetration for a selected binder, B.

---

[2] Felsinger, R., (1990). "Effect of Co-Mingled Polymer on Rheology of LDPE Modified Asphalt," Unpublished Report.

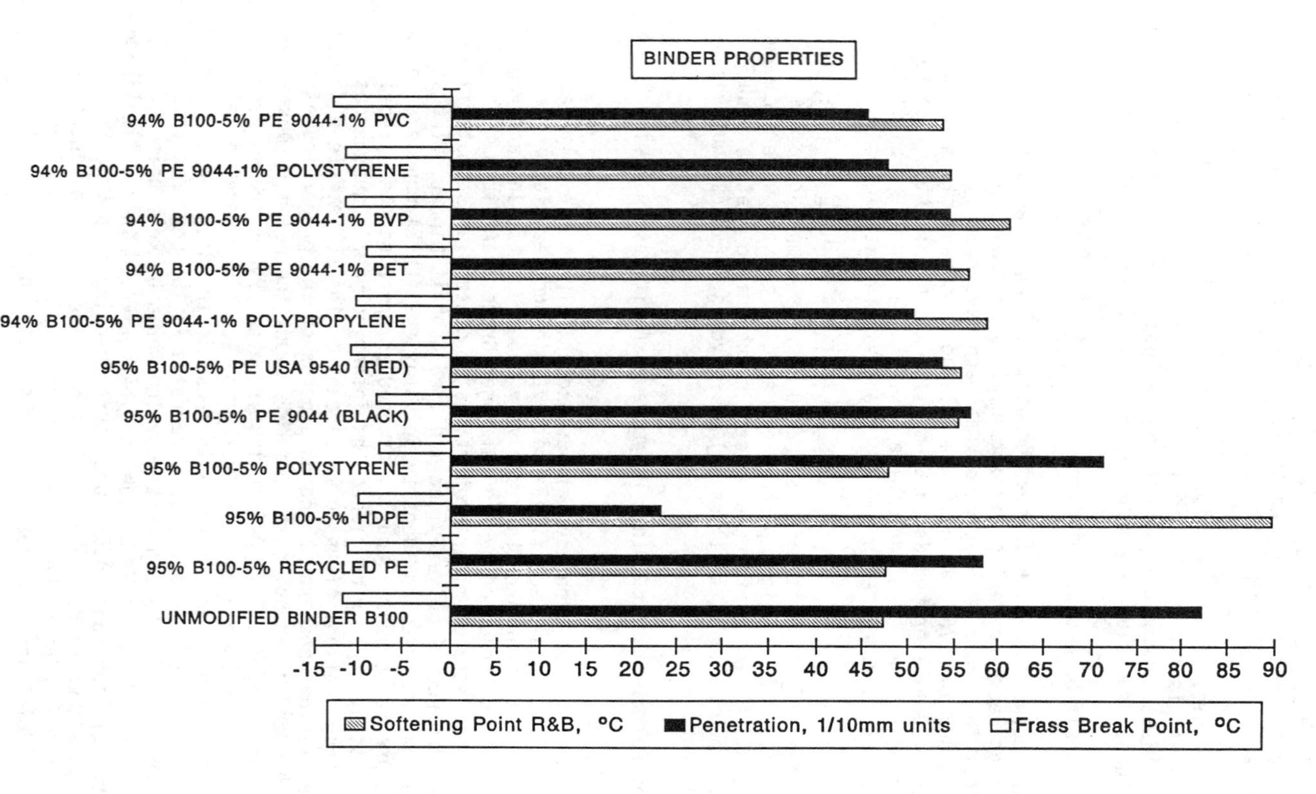

Figure 1. Summary of Binder Property Changes as a Result of Recycled Polyethylene (PE) Together with (Co-mingled with) other Polymers

Based on the Felsigner study, the Felsinger Group believes that they can produce their NOVOPHALT with a wide variety of recycled plastics from a wide range of suppliers, including from industrial and post-residential recyclers. They do, however, require that the recycled plastic be free of metal (because of milling difficulties and possible acceleration of aging) and paper (because hot mix asphalt concrete containing paper pulp may deteriorate faster when exposed to moisture).

ENGINEERING PROPERTIES OF SELECTED ASPHALT CONCRETE MIXTURES WITH LDPE-MODIFIED BINDER

Engineering Mixture Properties Selected for This Evaluation

Engineering mixture properties which can be used to predict the primary modes of distress were selected for mixture evaluation. The mixture properties measured were:

From Uniaxial Compressive Creep Testing

Strain at 3,600 seconds of loading, ϵ_p
Slope (log-log) of the steady state portion or the creep versus time of loading plot
Time to development of tertiary creep
Creep modulus at 3,600 seconds
Total resilient strain, ϵ_{rt}

From Indirect Tensile Testing, IDT

Diametrial resilient modulus
IDT strain at failure
Slope of the log-log IDT strain versus time of loading plot, m_t
IDT fatigue

From Controlled Displacement Fracture Propagation

J-integral
Paris law fracture parameters A and n
Fracture healing index

The process of creep in particulate media such as hot mix asphalt concrete has, on occasion, been explained as a rate process. The basis for the rate process theory is that atoms, molecules and particles participating in a time dependent flow process are constrained from movement relative to adjacent equilibrium positions. This displacement of flow units to new positions requires the introduction of activation energy of sufficient magnitude to surmount the barrier. Mitchell (4) explains that the rate process in particulate media is influenced by a number of factors as explained by the equation:

$$\dot{\epsilon} = 2X \frac{kT}{h} [-\frac{\Delta F}{RT}] \sinh [\frac{f\lambda}{2kT}] \quad (1)$$

where F is activation energy, T is absolute temperature, (°K), k is the Boltzman constant, h is Plank's constant, f is force, λ represents the distance between successive equilibrium positions, X represents the proportion of successful barrier crossings and R is the universal gas constant.

A schematic representation of the influence of creep stress intensity on creep rate at some selected time after stress application is presented in Figure 2. At low stress, creep rates are small and of little practical importance. At stresses approaching the strength of the material, the strain rates become very large and represent the onset of failure. In Figure 2 and in equation 1, it is apparent that the creep response of any particulate material, such as asphalt concrete, is not necessarily linear. If the stress state in the field (creep stress intensity) is one that pushes the strain rate into the region near failure, then assumptions of linearity are not correct.

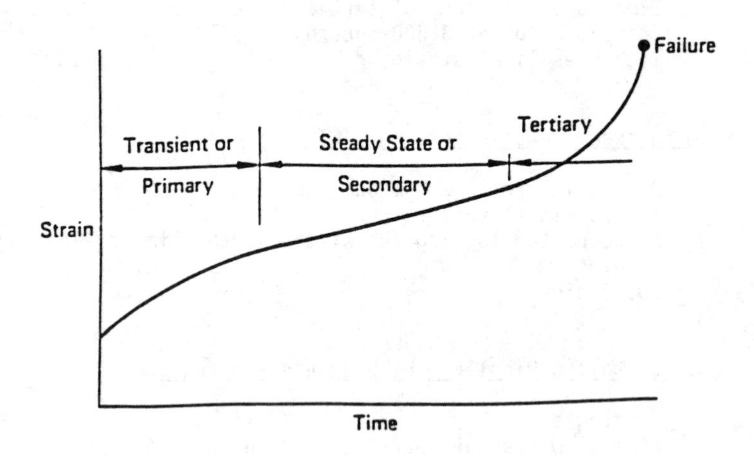

Figure 2. Stages of Creep (After Mitchell (4))

This point is very important because linear viscoelastic response of asphalt concrete mixtures under field loading conditions has often been assumed.

Another popular generalized form used to illustrate the various stages of creep is illustrated in Figure 3. In this figure the creep strain, for a given stress level, is plotted versus time, and the creep response is divided into three regions. The first region is primary creep where the rate of deformation increases rapidly. The second is steady state creep where the deformation rate is constant. The third is the region of failure, in which the deformation again increases rapidly.

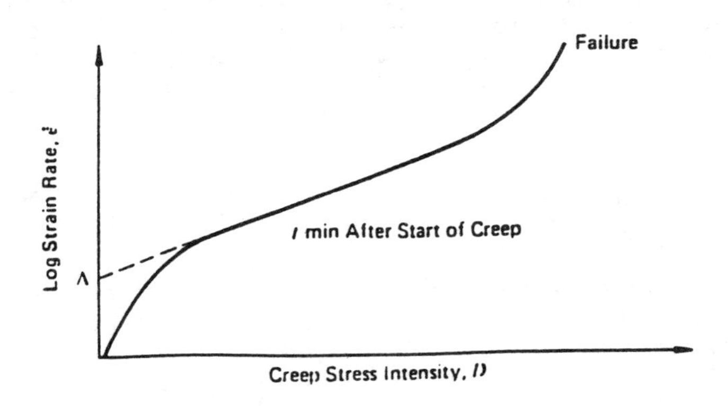

Figure 3. Influence of Creep Stress Intensity on Creep Rate (After Mitchell (4))

Obviously, the effects of aggregate gradation, aggregate type and mixture volume and void characteristics influence the creep response as identified by Hills (5) and Little and Youssef (6). The deformation potential of an asphalt concrete mixture is dependent on temperature of testing and stress state. The mixture creep response under a stress state that is typical of that actually induced in the pavement and at a temperature which is representative of that which occurs in the pavement is a reasonable predictor of permanent deformation susceptibility. And, although it may be difficult to precisely predict the amount of rutting which will occur under dynamic traffic loading, certain parameters from the creep strain versus time of loading plot are acceptable predictors or relative resistance to rutting among candidate mixtures. This is, however, true only when the creep load is applied at a temperature representative of field conditions and at a representative in situ stress state.

Little and Youssef (6) identified three parameters to be used to evaluate the creep versus time of loading response: the total creep strain at the end of one-hour or 3,600 seconds of loading, ϵ_p; the log-log slope of the steady state portion of the creep versus time of loading plot, m; and the creep stiffness at one-hour loading.

Little and Youssef (6) identified these parameters as important based on the characteristic shape of the creep strain versus time of loading plots of over 150 different mixtures tested over a range of stress levels. This research demonstrated that the majority of mixtures tested began to experience tertiary creep when they reach creep strains of in excess of 0.8 percent after one hour of creep loading. This research also established a correlation between the log-log slope of the steady state portion of the compressive creep strain versus time of loading plot and the onset of tertiary creep. Based on this relationship between the slope of the steady state region and critical strain at one hour of loading, it was possible to identify levels of ϵ_p and the slope of the steady state region, m, associated with various levels of permanent deformation potential. These criteria are summarized in Tables 1 and 2.

Table 1. Strain at One-Hour Creep Loading and Slope of Steady State Creep Curve Required to Reduce Rutting Potential to Very Low Level

Total Strain at One-Hour of Loading, %	Slope of Steady State Creep Curve					
	< 0.17	< 0.20	<0.25	< 0.30	< 0.35	< 0.40
< 0.25	IV[2]	IV[2]	IV[2]	IV[2]	IV[2]	III
< 0.40	IV[2]	IV[2]	IV[2]	III[2]	III[2]	III[2]
< 0.50	IV[2]	IV[2]	III[2]	III[2]	III[2]	II
< 0.80	III[2]	III[2]	II	II	II	II
< 1.0	I	I	I	I	I[1]	
< 1.2	I[1]	I[1]	I[1]			

Notes:

I - Low traffic intensity: $<10^5$ ESALs

II - Moderate traffic intensity: Between 10^5 and 5×10^5 ESALs

III - Heavy traffic intensity: Between 5×10^5 and 10^6 ESALs

IV - Very heavy traffic intensity: $>10^6$ ESALs

1. Must also have $\epsilon_p < 0.8\%$ at 1,800 seconds of creep loading
2. Should also meet the following criteria: $\epsilon_\pi + \epsilon_p < 0.5\ \epsilon_{qu}$

Table 2. Creep Stiffness Criteria at One-Hour Creep Loading

Level of Rut Resistance	Traffic Intensity Level	Required Minimum Creep Stiffness, psi, for Test Constant Stress Level of:		
		30 psi	50 psi	70 psi
Highly Rut Resistant	IV	15,000	17,500	22,500
	III	7,000	10,000	14,000
	II	5,000	6,500	8,750
	I	3,000	4,000	6,000
Moderately Rut Resistant	IV	7,500	10,000	14,000
	III	5,000	7,250	10,000
	II	3,500	6,000	7,500
	I	2,500	3,000	4,000

1000 psi - 6.895 MN/m^2

It is interesting that these criteria of creep stiffness at the end of one hour of loading are in acceptable agreement with those developed by other researches and through other studies, i.e., Sousa et al. (7), Mahboub and Little (8), Von Quintus et al. (9), Viljoen, et al. (10), Finn et al. (11) and Kronfuss et al. (12).

In addition to the criteria associated with the creep test, Von Quintus et al. (9) established the criteria that the sum of the total resilient strain from the dynamic compressive resilient modulus test (ASTM D 3497), ϵ_{rt}, and the creep strain at one hour of loading, ϵ_p, should not exceed one-half of the strain at failure in the compressive strength test, $0.5\epsilon_{qu}$ (AASHTO T 167). This is based on the evaluation of stress - strain plots from the unconfined compressive strength test which demonstrate that strain softening occurs at approximately $0.5\epsilon_{qu}$. This was substantiated based on unconfined compression testing on over 100 mixtures by Little and Youssef (6).

All compression testing was performed under realistic stress conditions and at a test temperature of 104°F (40°C). A realistic stress state in the uniaxial compressive test for most pavement conditions is between 40 and 70 psi (276 and 483 kN/m^2) as determined by Little and Youssef (6). This is based on a layered elastic analysis using realistic tire contact pressures and surface shear conditions and employing the Mohr-Coulomb failure theory which identifies the critical shear stress as that which occurs at the lowest level of normal stress developed within the asphalt pavement layer.

The diametral resilient modulus and indirect tensile strain at failure testing was performed to identify the susceptibility of the mixtures to fatigue cracking in accordance with the approach advocated by the Asphalt Aggregate Mixture Analysis System (AAMAS). In this approach the locus of the indirect tensile strain at failure and the diametral resilient modulus are plotted and compared to a control, standard dense graded mixture as shown in Figure 4. If the loci of points plot above the standard mix line, the mix in question possesses better than standard fatigue properties whereas if the loci are below the line the mix is inferior in terms of fatigue properties to the standard mix. This approach, Figure 4, can be used to rank the relative fatigue cracking potential for candidate mixtures.

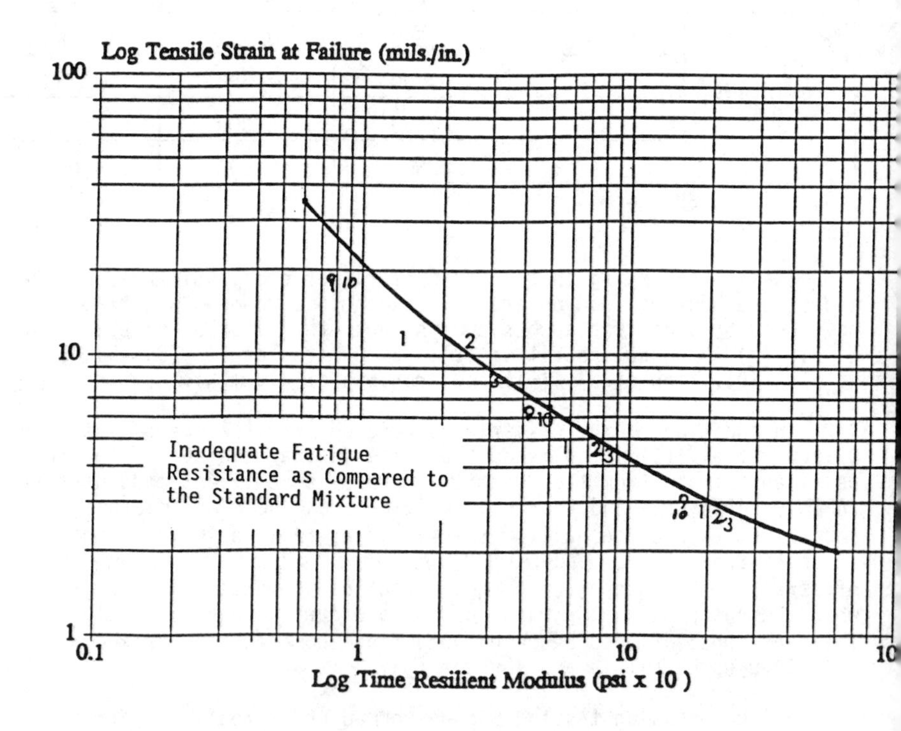

Figure 4. Minimum Tensile Failure Strains Required for the Mix as a Function of Resilient Modulus (1000 psi = 6.895 MN/m^2)

The indirect tensile creep response was selected as an indicator of the sensitivity of the mixture to fracture due to temperature shock. This slope parameter has been shown to be related to fracture through the laws of fracture mechanics and through empirical correlations to thermal cracking in the pavement. In terms of fracture mechanics, the Paris equation parameter n, which explains the rate of crack growth as a function of the stress intensity or J-integral (energy of crack extension in a visco-plastic medium) is inversely related to the slope of the indirect tensile creep curve m_t as:

$$n = \frac{2}{m_t} \tag{2}$$

The general form of the Paris fracture equation is:

$$\frac{da}{dN} = A\, J_{1c}^{\ n} \tag{3}$$

where A and n are material constants which identify the rate of change in crack length (crack growth) da per cycle dN, and J_{1c} is the change in the energy which drives the crack in the visco-plastic material. Schapery's (13) landmark investigation of linear viscoelastic fracture identified that a theoretical relationship exists between m_t and n and between m_t and A. Extensive testing by Little and Youssef (6) have empirically verified the relationship between m_t and the fracture parameters.

Results of Compressive Testing

Results of uniaxial compressive testing are summarized in Table 3 for 12 selected dense-graded asphalt concrete mixtures and for 4 stone mastic asphalt (SMA) mixtures. The difference between the SMA and dense-graded mix is that, whereas the dense-graded mix derives its shear strength from a well-graded assortment of high quality (crushed, rough-textured, hard and durable) aggregate particles, the SMA is a gap-graded, coarser mix. The gap gradation allows a significantly larger portion of coarse aggregate (larger than 2 mm) than does the dense-graded mix (i.e., 70-80 percent versus 40-70 percent). In the SMA, stone-to-stone interaction among the coarse fraction is supposed to provide stability. The relatively large void volume among the coarse stones is filled with mastic. The net result is a binder-rich, mastic-rich, stable mix. SMA mixtures typically use fibers, polymer modification or both to provide stability of the binder-rich mix during placement and in-service.

All of the mixtures in Table 3 incorporate the optimum binder content and were compacted to a target air void content of approximately between 3 and 5 percent.

Columns 4 through 7 present the total strain at 3,600 seconds of loading, ϵ_p; the log-log slope of the steady state creep region, m; time to rupture, t_r; and creep modulus, E_c, respectively, for all mixtures. Within each selected mixture category (type), it is evident that the addition of recycled LDPE in the range of 4 to 6 percent reduces rutting potential as is reflected by a significant reduction in ϵ_p and m and a significant increase in t_r and E_c.

Table 3. Summary of Pertinent Uniaxial Compression Test Data and Parameters for Selected Mixtures

Aggregate	Binder	Air Voids, %	ϵ_p, in/in	M	t_r, sec.	E_c, psi	ϵ_{rt}, in/in	SSSF	q_u, psi
100% Crushed	AC-10	3.2	0.0050	0.23	7,000	12,000	0.00030	1.6	300
	AC-10+4.3% LDPE	3.7	0.0035	0.17	>10,000	17,143	0.00025	2.2	350
	AC-10+6.0% LDPE	3.4	0.0025	0.15	>10,000	24,000	0.00025	3.2	410
90% Crushed 10% Natural Sand	AC-10	3.8	0.0085	0.32	3,200	7,060	0.00035	0.9	260
	AC-10+5% LDPE	4.2	0.0065	0.25	6,000	9,231	0.00030	1.2	310
80% Crushed 20% Natural Sand	AC-10	3.3	0.0095	0.42	2,800	6,316	0.00040	0.8	225
	AC-10+5% LDPE	3.4	0.0070	0.22	5,500	8.571	0.00037	1.1	280
100% Rounded River Gravel (RG)	AC-20	4.2	0.0140	0.40	2,000	4,286	0.0009	-	-
80% RG+20% Crushed	AC-20	4.4	0.0120	0.30	3,000	5,000	0.00075	0.7	120
80% RG+20% Crushed	AC-20+5% LDPE	3.9	0.0085	0.24	5,000	7,060	0.00059	0.9	155
100% Crushed Granite	AC-20	5.1	0.0065	0.30	4,000	9,230	0.00029	-	-
	AC-20+5% LDPE	5.0	0.0045	0.17	20,000	13,333	0.00020	-	-
Stone Mastic Mixture (SMA)	AC-30	3.0	0.0150	0.35	2,000	4,000	0.0012	-	-
0.3% Fiber (Georgia Granite)	AC-30+5% LDPE	3.0	0.0080	0.20	>3,600	7,500	0.0010	-	-
SMA with Crushed Gravel (Colorado)	AC-10	2.8	0.0080	0.29	3,600	7,500	0.0008	-	-
0.3% Fiber	AC-10+LDPE	3.0	0.0065	0.20	>3,600	9,230	0.0006	-	-

1 in = 25.4 mm 1000 psi = 6.895 MN/m^2

The addition of 6 percent recycled LDPE to the 100 percent crushed limestone mix doubles the E_c. This may represent an increase in stiffness and rigidity of the mix that is too severe for many applications, especially those in which load-induced fatigue cracking, thermal shock, and/or thermal fatigue are of concern. Perhaps a lower percentage of LDPE (i.e., 4.3 percent) is better suited for modification of this mix.

The 80 percent crushed limestone and 20 percent natural (uncrushed) sand mix represents a mix typical of many used in Texas. Without LDPE modification the ϵ_p (0.0095 in./in.) exceeds the maximum guidelines for permanent deformation resistance (0.008 in./in.) set forth in Table 1 and, consequently, tertiary creep is reached in only 2,800 seconds. The addition of 5 percent recycled LDPE reduces ϵ_p to 0.0070 in./in. and hence extends the time to tertiary creep to well beyond the typical creep testing time of 3,600 seconds. According to historically accepted creep modulus criteria, this modification increased the modulus from a marginal or borderline value of 6,300 psi (43.4 MN/m^2) to an acceptable value of 8,600 psi (59.3 MN/m^2). Similar results are evident in the 80 percent river gravel/20 percent crushed limestone mix.

In each case for the dense-graded mixes in Table 3, the creep modulus is substantially increased, indicating greater stability an a greater factor of safety against entering into tertiary creep, and the slope, m, is significantly reduced, indicating a significantly lower rate of creep in this region.

Another criterion by which to judge deformation (creep) and stability (resistance to shear failure) is to evaluate the strain softening safety factor which is defined as:

$$SSSF = \frac{0.5\epsilon_{qu}}{\epsilon_p + \epsilon_{rt}} \tag{4}$$

As has been discussed, when the SSSF ratio is greater than one, the stress-strain region associated with strain softening should not be entered. For each dense-graded mix the addition of recycled LDPE significantly increases its SSSF as well as the unconfined compressive strength, q_u.

Several field projects have allowed a comparison between the performance of LDPE-modified asphalt and unmodified asphalt (control) on otherwise identical mixtures. A particularly interesting demonstration project on I-75 in Georgia allowed a direct comparison of rutting potential between the control mix and the modified mix. The mixture placed included recycled asphalt pavement (RAP) at from 15 percent to 25 percent by weight of the mix in addition to recycled plastics (LDPE) by the Novophalt process. Plant mix samples were evaluated by the Georgia Department of Transportation's loaded wheel test (LWT) approach.

The LWT results demonstrated that the LDPE-modified mix resulted in 50 percent to 90 percent reduction in high temperature deformation and a predicted pavement life increase of more than 50 percent greater than that of the control (unmodified) mix (Georgia Department of Transportation, 1991).[3]

Results of Indirect Tensile Testing

Results of indirect tensile (IDT) testing are summarized in Table 4. These results will be discussed in two categories: load-induced fatigue prediction based on IDT strain at failure and diametral resilient modulus as a function of temperature. The loci of IDT strain at failure and diametral resilient moduli for the selected mixtures are plotted in Figure 4 and can be compared against the standard mix. From this figure it can be seen that the load-induced fatigue properties of the LDPE modified mixes are essentially not different from those of the unmodified control mix within each group. The IDT strain at failure data and resilient moduli data in Table 4 were recorded on samples that were aged in accordance with the procedures described in NCHRP Report 338 (9).

In support of the data in Table 4 and summarized in Figure 4, Table 5 presents the results of constant stress IDT repeated load fatigue testing for mixtures of river gravel aggregate and three different binders: AC-10, AC-10 modified with 5 percent LDPE and AC-20 grade asphalt. In each case the source of the asphalt was a Texaco refinery. Two levels of aging were considered: unaged and aged for 21 days at 140°F (60°C) in an oven. From this testing the following information is gleaned:

1. In the IDT controlled stress mode of loading, the stiffening effect of LDPE improves fatigue life at 77°F (25°C). This is probably because the elastic component of stiffness is increased due to the addition of LDPE. The net result is a reduction in delayed elastic and viscous deformation and a reduction in cumulative damage during loading. Such a response is logical for an asphalt concrete layer as part of a pavement system in a controlled stress mode.

2. The effect of aging is pronounced at 77°F (25°C). The stiffening effect of aging improves fatigue life in the controlled stress mode. The effect of aging on the LDPE-modified samples is more pronounced (actually more favorable in the controlled stress mode) not because aging is more deleterious but because the synergistic interaction of aging and the LDPE modification produces a stiffer mix than does a single mode of stiffening (i.e., modification alone or aging alone). Petersen et al. (14) demonstrated that the increase in carbonyl, carboxylic or sulfoxide functional groups due to oxidative aging was no greater in LDPE-modified asphalts than for modified asphalts.

---

[3] Georgia Department of Transportation, (1991). Unpublished Report on Research Project 9003: "Comparison of Performance of Standard Asphalt Mixes to Mixes Using Asphalt Concrete Modifiers."

Table 4. Summary of Pertinent Indirect Tensile Test Data for Selected Mixture

Aggregate	Binder	Mix No.	Diamtral Resilient Modulus, psi			Indirect Tensile Failure Strain, in/in		
			41°F	77°F	104°F	41°F	77°F	104°F
100% Crushed	AC-10	1	1,995,000	650,000	175,000	0.0030	0.0050	0.0100
	AC-10+4.3% LDPE	2	2,100,000	795,000	250,000	0.0029	0.0048	0.0100
	AC-10+6.0% LDPE	3	2,200,000	810,000	260,000	0.0025	0.0047	0.0090
90% Crushed 10% Natural Sand	AC-10 AC-10+5% LDPE	4 5	-	-	160,000 190,000	-	-	-
80% Crushed 20% Natural Sand	AC-10 AC-10+5% LDPE	6 7	-	-	140,000 175,000	-	-	-
100% Rounded River Gravel (RG)	AC-20	8	1,700,000	320,000	45,000	-	-	-
80% RG+20% Crushed	AC-20	9	1,800,000	390,000	50,000	0.0034	0.0060	0.019
80% RG 20% Crushed	AC-20 +5% LDPE	10	1,750,000	450,000	72,000	0.0028	0.0062	0.019
100% Crushed Granite	AC-20 AC-20+5% LDPE	11 12	2,200,000 2,500,000	650,000 675,000	195,000 210,000	-	-	-
Stone Mastic Mixture (SMA)	AC-30	13	2,000,000	300,000	50,000	-	0.0079	-
0.3% Fiber (Georgia Granite)	AC-30+5% LDPE	14	2,000,000	350,000	90,000		0.0075	

1000 psi = 6.895 MN/m^2 °C = 5(°F - 32)/9

Table 5. Summary of the Results of IDT Repeated Load Fatigue Testing of River Gravel Mixtures Bound with Texaco Asphalt: AC-10 + LDPE and AC-20

Type of Asphalt	Stress (psi)	Temp. °F	Cycles to Failure w/o aging	Aged Samples
AC-10	30	77	3,391	-
	40	77	1,316	3,231
	50	77	627	1,824
	70	77	-	569
	188	33	6,327	5,513
	200	33	4,027	3,283
AC-10 + 5% LDPE	30	77	16,408	-
	40	77	5,967	27,819
	50	77	2,898	11,742
	70	77	-	3,332
	188	33	13,970	9,048
	200	33	6,481	6,143
AC-20	30	77	8,507	-
	40	77	4,946	14,144
	50	77	2,584	7,108
	70	77	-	2,476

1000 psi = 6.895 MN/m^2
1°C = 5(°F-32)/9

3. At 33°F (0°) and at large strains, the stiffening effect induced by aging reduces fatigue life. This is because the large strains induced produce fracture at this relatively low temperature, and the viscous flow component that is needed to release energy without the formation of new fracture faces is absent.

Another indication of the synergistic effects of aging on the fatigue an fracture potential of LDPE-modified mixtures is shown in Figure 5, which presents IDT creep data at 77°F (25°C). As previously discussed, the slope of the creep relationship is directly and keenly related to both thermal fracture potential and load-induced fatigue. The log-log slope of the unaged, unmodified mixture is approximately 0.5 while the slope of the unaged, modified mix is 0.32. The corresponding slopes of the aged mixtures are 0.35 and 0.27, respectively. Thus, the effect of the LDPE is to reduce the slope when compared to the unmodified mix, and the synergistic effect of aging and LDPE-modification is to reduce the slope to the lowest level among the mixes illustrated in Figure 5.

It should also be noted in Figure 5 that the magnitude of strain induced is substantially lower for the LDPE-modified mixes than for

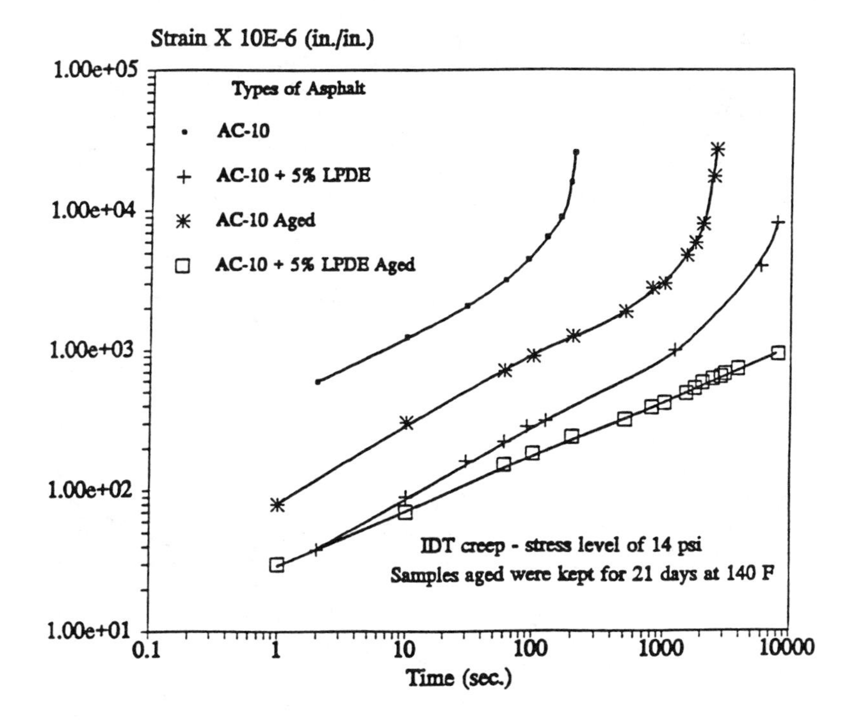

Figure 5. IDT Creep Results at Applied Stress Level of 14 psi for River Gravel Mixtures Bound with Texaco Asphalts: AC-10, AC-10 + LDPE. (Each Data Point Represents the Average of Three Replicates Samples)

the unmodified; and, correspondingly, the time to tertiary creep is substantially longer for the LDPE-modified mixes than for the unmodified mixes. Navarro and Kennedy (15) identified a strong correlation between the time to rupture from the IDT creep test to controlled stress fatigue failure at moderate and high temperature (77°F (25°C) and 104°F (40°C), respectively), with longer times to rupture correlating with longer fatigue lives.

It is interesting to note that tertiary creep initiates at approximately one percent strain which is similar to what has been shown to occur in the compressive creep mode.

Fracture due to controlled displacement is quite different from fatigue due to the controlled stress mode of loading. In controlled displacement, damage is usually mitigated by a material that can dissipate energy at the crack tip through flow or deformation. Hence, more deformable materials often perform better than stiffer materials in which stored energy results in crack growth.

A notched beam was tested in direct tensile, controlled deformation to evaluate resistance to reflection cracking. In this test the rate of crack growth was measured as a function of the energy at the crack tip as measured by the J*-integral (a modified version of the J-integral for cyclic loading). Figure 6 presents the results of controlled displacement fracture for river gravel mixtures with Texaco AC-20 grade asphalt or Texaco AC-5 grade asphalt with modification.

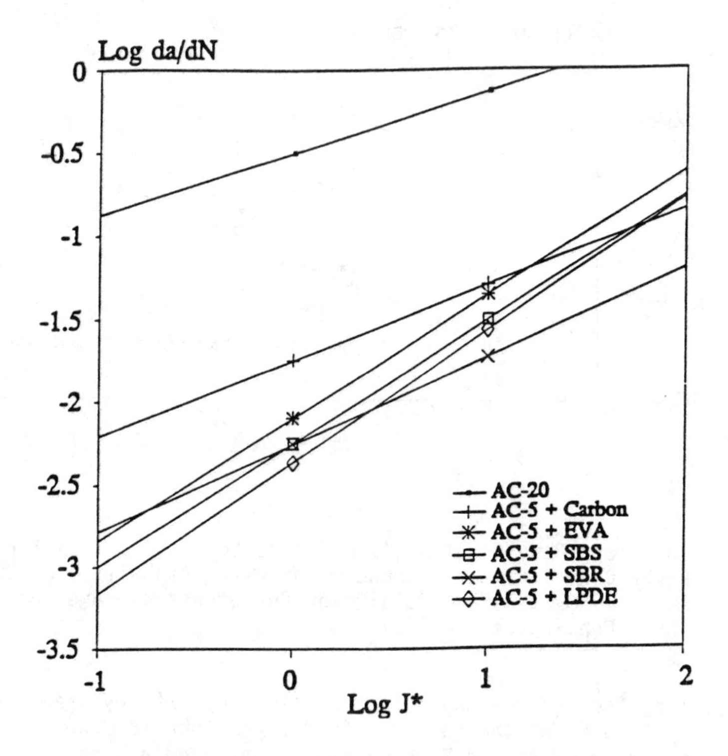

Figure 6. Results of Controlled Displacement Crack Propagation Testing of Selected Mixtures Bound with Texaco AC-5 Asphalt and Various Additives: Carbon Black, Ethylene Vinyl Acetate, Styrene Butadiene Styrene, Styrene Butadiene Rubber and LDPE (Novophalt)

Based on the intercept and slope of the da/dN v. J-integral relationship, the LDPE-modified AC-5 mix possesses fracture properties that are consistent with the other polymer modifiers and far superior to the AC-20 mix. The testing presented in Figure 6 was performed at 60°F (15°C).

When rest periods are introduced during the controlled strain test using the notched beam, the number of cycles to failure or the number of cycles required to cause the crack to propagate through the test beam is significantly increased. The reasons for this "healing effect" are discussed by Kim and Little (16). The healing effect occurs in the microcracks that proceed the macrocrack, and as a result of the microcrack healing, the propagation of the macrocrack is significantly delayed. Balbissi and Little (17) have demonstrated that the effects of the healing phenomenon on fatigue life can be substantial and may increase fatigue life by orders of magnitude. Figures 7 and 8 compare the healing index, a measure of the energy required to propagate a crack after a rest period as compared to the energy required before the rest period, to the ratio of length of rest period to number of loading cycles. For a California Valley asphalt (AC-10), the addition of recycled LDPE improves the healing index significantly for testing performed at 77°F (25°C). In fact regression equations were developed to relate healing index to the ratio of rest period and number or rest periods and to strain amplitude (per loading cycle). These relationships for the mixtures of crushed granite and California Valley AC-10 asphalt cement are:

Unmodified

$$\log HI = -4.684 + 0.321 \log (RP/NP) - 0.836 \log (SA) \tag{5}$$

Modified

$$\log HI = -2.913 + 0.324 \log (RP/NP) - 1.102 \log (SA) \tag{6}$$

where RP is length of rest period in minutes, NP is number of loading cycles and SA is strain amplitude.

The cumulative effects of asphalt modification with recycled LDPE on fatigue fracture life appears to be beneficial for controlled stress, load-induced fatigue and for controlled strain fracture fatigue at intermediate temperatures. More low temperature testing is required to fully evaluate the effects at low temperatures. Although LDPE modification may marginally reduce fracture fatigue life and increase thermal fatigue potential at low temperature, Goodrich (18) has demonstrated that low temperature properties of polymer-modified asphalt are dominated by the properties of the asphalt cement. Thus, it should be possible to engineer a LDPE-asphalt blend to reduce high-temperature deformation and resist low-temperature fracture.

CONCLUSIONS

Recycled low density polyethylene can be very effective in reducing the permanent deformation or rutting potential of asphalt concrete mixtures if the LDPE is mixed with the asphalt in the proper manner. A high shear blending process at elevated temperature was used for recycled LDPE dispersion into the asphalt cement in the study.

The controlled stress fracture fatigue properties of the mixture are enhanced at moderate pavement temperatures, i.e., at approximately 77°F (25°C). However, at low temperatures, the LDPE-modified mix may

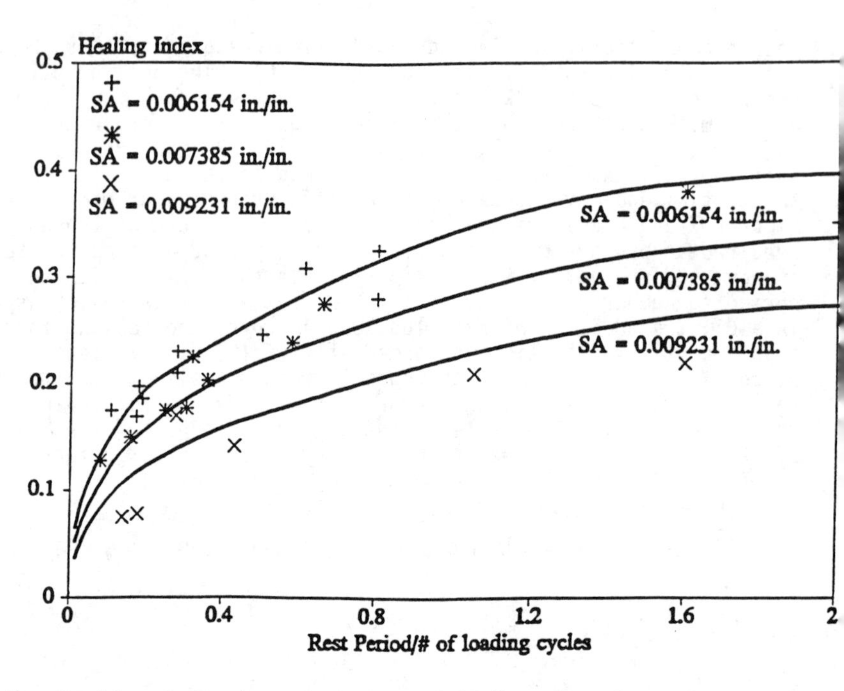

Figure 7. Healing Index as a Function of RP/# and SA for California Valley Asphalt Modified with 5 Percent LDPE

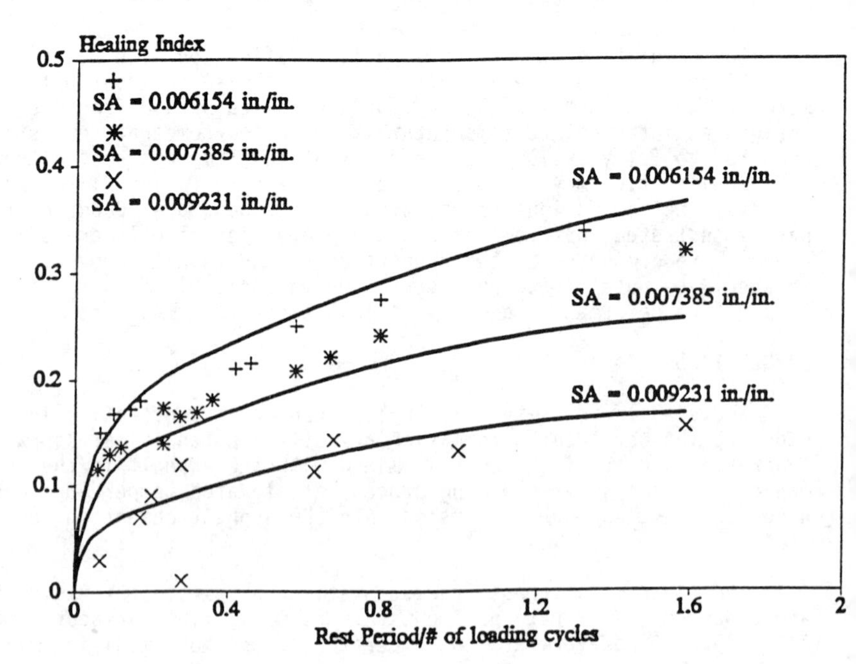

Figure 8. Healing Index as a Function of RP/# and SA for California Valley (virgin) Asphalt

be slightly more susceptible to fracture fatigue in either a controlled stress or a controlled strain mode. The low temperature fracture and fatigue potential is controlled mainly by the properties of the binder.

Small to moderate levels of co-mingling with other polymers (up to approximately 15 percent) appears to be acceptable in terms of the influence of the LDPE on the rheological properties of the binder.

Low density polyethylene modification of asphalt concrete mixtures appears to an especially attractive consideration in the warm climates of the Southern United States.

REFERENCES

1. "Guidelines for Ferrous Scrap, Nonferrous Scrap, Paper Stock and Plastic Scrap," (1991). Scrap Specifications Circular, Institute of Scrap Recycling Industries, Inc., Washington, D. C.

2. Gardner, J., (1992). "Plastics Recycling on Rise, EPA Report Says," Plastics News, Vol. 4, No. 23.

3. Mitchell, J. K., (1976). Fundamentals of Soil Behavior, John Wiley and Sons, 1976.

4. Hills, J. F., (1973). "The Creep of Asphalt Mixes," Journal of the Institute of Petroleum, Vol. 59.

5. Little, D. N. and Youssef, H., (1992). "Improved ACP Mixture Design: Development and Verification," Research Report 1170-17, Texas Transportation Institute, Texas A&M University.

6. Sousa, J. B., Harvey, J., Painter, L., Deacon, J. A. and Monismith, C. L., (1991). "Evaluation of Laboratory Procedures for Compacting Asphalt Aggregate Mixtures," Institute of Transportation Studies, University of California at Berkeley.

7. Mahboub, K. and Little, D. N., (1987). "Improved Asphalt Concrete Mixture Design," Research Report 2474-1F, Texas Transportation Institute, Texas A&M University.

8. Von Quintus, H. L., Scherocman, J. A., Hughes, C. S. and Kennedy, T. W., (1991) "Asphalt Aggregate Mixture Analysis System (AAMAS)," NCHRP Report 338.

9. Viljoen, A. W. and Meadows, K., (1981). "The Creep Test - A Mix Design Tool to Rank Asphalt Mixes in Terms of Their Resistance to Permanent Deformation Under Heavy Traffic," National Institute of Road Research, Pretoria, South Africa.

10. Finn, F., Monismith, C. L. and Markevich, N. J., (1983). "Pavement Performance and Asphalt Concrete Mix Design," Proceedings, Association of Asphalt Paving Technologists, Vol. 52.

11. Kronfuss, R., Krzemien, R., Nievelt, G. and Putz, P., (1984). "Verformungsfestigkjeit von Asphalten Ermittlung in Driechtest, Bundesministerium fur Bauten and Technik, Strassenforschung," Heft 240, Wien, Austria.

12. Schapery, R. A. (1984). "Correspondence Principles and a Generalized J-Integral for Large Deformation and Fracture Analysis of Viscoelastic Media," International Journal of Fracture, Vol. 25.

13. Petersen, C. (1988). "Evaluation of Oxidation Aging of LDPE-Modified Asphalt Cement," Western Research Institute.

14. Navarro, D. and Kennedy, T. W. (1975). "Fatigue and Repeated-Load Elastic Characteristics of Inservice Asphalt-Treated Materials," Research Report 183-2, Center for Highway Research, The University of Texas at Austin.

15. Kim, Y. R. and Little, D. N., (1990). "Chemical and Mechanical Evaluation of Healing Mechanism of Asphalt Concrete," Proceedings, Association of Asphalt Paving Technologists, Vol. 59.

16. Balbissi, A. H. and Little, D. N., (1983). "A Comparative Analysis of the Fracture and Fatigue Properties of Asphalt Concrete and Sulphlex," Ph.D. Dissertation, Texas A&M University, College Station, Texas.

17. Goodrich, J. L. (1991). "Asphaltic Binder Rheology, Asphalt Concrete Rheology and Asphalt Concrete Mix Properties," Proceedings, Association of Asphalt Paving Technologists, Vol. 60.

Papers Not Presented at Symposium (Publication Only)

Serji N. Amirkhanian[1]

UTILIZATION OF SCRAP TIRES IN FLEXIBLE PAVEMENTS - REVIEW OF EXISTING TECHNOLOGY

REFERENCE: Amirkhanian, S.N., **"Utilization of Scrap Tires in Flexible Pavements - Review of Existing Technology,"** Use of Waste Materials in Hot-Mix Asphalt, ASTM STP 1193, H. Fred Waller, Ed., American Society for Testing and Materials, Philadelphia, 1993.

ABSTRACT: In recent years, the increasing need for recycling has forced highway engineers to search for more innovative and economical methods of building and maintaining the nation's highways. There is an urgent need to find ways to recycle waste materials (e.g., tires) to reduce the amount of waste going into landfills each year. In this review of the existing technology, the feasibility of the use of scrap tires in flexible pavements was investigated. An extensive literature review was conducted. In addition, an extensive survey of all state and federal highway-related agencies was performed. Based on the literature review and surveys, several recommendations were submitted to the South Carolina Department of Highways and Public Transportation and Federal Highway Administration officials.

KEYWORDS: waste tires, flexible pavements, rubber, SAM, SAMI, dry and wet process

[1] Assistant Professor, Civil Engineering Department, Clemson University, Clemson, SC 29634-0911.

INTRODUCTION

In recent years, the increasing need for recycling has forced highway engineers to search for more innovative and economical methods of building and maintaining the nation's highways. As solid waste landfill space is becoming a problem throughout the nation (e.g., cost and environmental issues), there is an urgent need to find ways to recycle waste materials (e.g., tires) to reduce the amount of waste going into landfills each year.

Scrap tire recycling mandates are written into both the Intermodal Surface Transportation Efficiency Act (ISTEA) of 1991 and the Resource Conservation Recovery Act stating that federal-aid highway funding would be withheld from states if they do not comply with these Acts [1]. Section 1038 of the ISTEA (Use of Recycled Paving Material) addresses the use of scrap tires in asphaltic concrete mixtures and contains three primary requirements including: 1) the federal regulations regarding the use of scrap tires must be relaxed, 2) the performance, recycling, and environmental issues related to use of scrap tires in asphaltic concrete mixtures must be studied, and 3) each state must satisfy a minimum scrap tire utilization requirement starting 1994.

In the last three years, there has been an increase in state legislations' interest in the use of waste materials in pavements. For example, as of January 1989, only five states regulated disposal of used tires. However, by January 1991, thirty five states had laws or regulations addressing this issue [1]. However, due to limited use of rubber in asphaltic concrete pavements in the United States, the information on performance of rubberized asphaltic concrete mixtures is fragmented.

The South Carolina State Legislature, in an effort to reduce the amount of solid waste generated in the state, passed the South Carolina Solid Waste Policy and Management Act of 1991. This Act requires the South Carolina Department of Highways and Public Transportation (SCDHPT) to investigate the use of certain waste products (e.g., tires) in various aspects of highway construction. Because of this Act and environmental concerns, the SCDHPT proposed to initiate a study to investigate the feasibility of use of scrap tire rubber in asphaltic concrete mixtures. This paper reports on a portion of an on-going research project in this area.

BACKGROUND

There are approximately 240 million waste tires generated annually in the United States. Of these, 200 million are passenger car tires and 40 million are truck tires [2]. According to the industry figures, there are approximately 2 to 2.5 billion scrap tires currently on the ground in the United States [2].

A typical scrap tire (passenger car) weighs approximately 9 kilograms (20 pounds) and will provide approximately 60% rubber, 20% steel and 20% fiber and other waste products [3]. Currently, the paving industry uses 1 to 2 million tires per year. Each metric ton of Hot Mix Asphalt (HMA) which contains rubber can utilize 2 to 6 tires [4]. For example, if 50 million scrap tires are going to be utilized in asphaltic concrete mixtures, then 8 to 25 million metric tons of HMA would require modifications.

There are many factors to consider when a state agency uses rubber in its pavements including cost, specifications, type of equipment to be used, expertise of the contractor, potential recyclability of materials, etc. There are many advantages for using rubber in asphaltic concrete mixtures. Some of the potential (or reported) benefits of mixtures containing rubber include:

1. thinner lift,
2. increased pavement life,
3. retarded reflection cracking,
4. decreased traffic noise,
5. reduced maintenance costs, and
6. decreased pollution and increased environmental quality.

There are many issues and problems associated with the use of tires in asphaltic concrete pavements that must be researched and analyzed. Some of the issues and problems include:

1. High initial costs: Some highway agencies and/or private firms manufacturing the scrap tires claim an increase of approximately 25% to over 200% in the cost of the pavement,

2. Lifecycle economics: There are claims of doubling the life of conventional pavements by using these products; however, there has not been a comprehensive study to justify the high initial costs and to determine the lifecycle economics of this type of pavement,

3. Lack of product specifications by ASTM or any other agency,

4. The uniformity of scrap tire rubber could be a problem and must be studied in great detail,

5. The recyclability of the pavement containing scrap tires must be investigated and analyzed,

6. There are some environmental concerns over the use of tires in a pavement, and

7. There are some concerns over necessary modifications made to the asphalt plants or equipment used for a typical paving operation.

METHODS OF UTILIZATION

Two major processes used in designing and constructing an asphalt-rubber mixture are referred to as "wet" and "dry" processes. In the wet process, rubber is added to asphalt cement. In the dry process, crumbed rubber is used as a portion of the aggregate. For example, in one of the wet processes used in some parts of the country, approximately 5% to 10% of rubber (by weight of the asphalt cement) and in one of the dry processes approximately 3% of rubber (by weight of the total mixture) are added to the asphaltic concrete mixtures. There have been many projects constructed with these two methods; however, the long-term performance and life-cycle cost comparisons of the two methods (wet vs. dry) have not been researched in depth.

The first application that used the wet process is referred to as the McDonald process. Charles McDonald (Materials Engineer for the City of Phoenix, Arizona) with cooperation from a local asphalt company (Sahuaro Petroleum), blended crumb rubber with asphalt cement and used the mixture as a surface treatment. Arizona Department of Transportation placed its first Stress Absorbing Membrane (SAM) in 1968, its first Stress Absorbing Membrane Interlayer (SAMI) in 1972, and its first application of the use of asphalt rubber in HMA open graded friction course in 1975 [4].

The first application of the dry process is referred to as PlusRide process. This technology originated in the 1960's by the Swedish companies Skega AB and AB Vaegfoerbaettringar (ABV) and was patented as Rubit (the European trade name for this process). In 1978, it was patented in the United States as PlusRide and marketed by All Seasons Surfacing Corporation of Bellevue, Washington. Today, EnvirOtire, Inc. of Seattle, Washington, retains all patent rights and works with each contractor the terms of specific licensing agreements.

PROCESSING METHODS AND COST

There are four methods used for processing scrap tires into crumb rubber including: the cryogenic process, granulator process, micro-mill process, and crackermill process. The cryogenic process involves the submersion of a scrap tire in a bath of liquid nitrogen, providing a brittle tire that is crushed to specified size. This technique is very costly compared to other methods and some states do not allow the use of cryogenically crushed rubber due to the surface texture. The granulator process involves shearing apart the scrap tire rubber and cutting it with revolving steel plates at ambient temperatures. The size of produced particles can range from 9.5 millimeter sieve (3/8 inch) to 2.0 millimeter sieve (No. 10 sieve).

The micro-mill process reduces the rubber size to very finely ground particles. The particles can be reduced to sizes ranging from 425 micron sieve (No. 40 sieve) to 75 micron sieve (No. 200) sizes. The crackermill process (the most widely used method) uses rotating corrugated steel drums to tear apart scrap tires at ambient temperature. In this process, the scrap rubber must be pre-processed by shredding. The size of produced particles can be over a range of 4.75 millimeter sieve (No. 4) to 425 micron sieve (No. 40 sieve).

The cost of coarse and medium sized crumb rubber (above 425 micron (No. 40) sieve) is approximately 20 to 35 cents/kilogram (10 to 15 cents/pound) and the cost for fine ground rubber is over 50 cents/kilogram (23 cents/pound). Most crumb rubber is shipped in 23 to 27 kilogram bags (50 to 60 pounds). However, the rubber can be shipped in 900 kilogram (2000 pounds) containers which some of them are reusable.

OBJECTIVE

The major objective of this research effort was to study the feasibility of the use of waste rubber tires in asphaltic concrete mixtures. In order to achieve this objective the following steps were followed:

1. An extensive survey (e.g., questionnaire) of all state and federal highway-related agencies was conducted.

2. An extensive literature review of work performed in this area was conducted.

SURVEY RESULTS

A questionnaire was prepared and mailed to all state highway agencies and some of the federal highway-related agencies in the United States. The following state highway agencies responded to the questionnaire: Alaska, Arkansas, California, Colorado, Connecticut, District of Columbia, Florida, Georgia, Hawaii, Illinois, Iowa, Kentucky, Maine, Maryland, Michigan, Minnesota, Mississippi, Missouri, Nebraska, Nevada, New Hampshire, New Mexico, New Jersey, New York, Oregon, Pennsylvania, Rhode Island, South Carolina,

South Dakota, Tennessee, Texas, Vermont, and Wisconsin. The rate of response was 66%.

Nineteen states indicated that they were currently using or had constructed some experimental sections made with mixtures containing rubber. These states included: Alaska, Arkansas, California, Colorado, Connecticut, Florida, Georgia, Illinois, Iowa, Maine, Maryland, Michigan, Missouri, Nebraska, New Hampshire, Oregon, Pennsylvania, Texas, and Wisconsin. Most of these states indicated that the use of mixtures containing rubber is in its experimental stages.

The following is a summary of the results of the questionnaire provided by the participating agencies:

1. Approximately how many miles of new flexible pavement were constructed in your state this year?

   a. 10.0% 0-10 b. 38.0% 11-30 c. 14.0% 31-50
   d. 7.0% 51-100 e. 31.0% >150

2. Approximately how many miles of asphaltic concrete overlay were placed in your state this year?

   a. 10.0% 11-50 b. 7.0% 51-100 c. 10.0% 101-150
   d. 3.0% 201-250 e. 3.0% 251-300 f. 67% >300

3. Is there legislation (on books/pending) calling for a waste rubber recycling program in your state?

   38% yes 62% no

4. Please describe the parameters of waste rubber tire disposal in your state.

   * $2.00/new tire (charged to the user)
   * Landfills * Unknown * Recycling
   * $4-5/whole tire: for disposal

5. Is your organization presently using mixtures containing rubber for flexible pavement construction?

   14.0% yes 86.0% no

6. Is your organization presently using mixtures containing rubber for pavement repairs or overlays?

   41.0% yes 59.0% no

7. With regard to mixtures containing rubber, what general disadvantages have you experienced?

* Cold temperature cracking
* Shorter compaction period
* Constructability
* Increased mix temperature

8. Have you experienced any problems in the mixing of mixtures containing rubber?

51.0% yes (if yes, please describe) 49.0% no

* Cost
* Limited supply of rubber
* Excessive plant emissions
* Dry process had loss of fines
* Increased temperatures and viscosity
* Difficult to introduce it into machine
* Rubber's tendency to stay in silo and the drum drier

9. Have you experienced any problems in the placing of mixtures containing rubber?

38.0% yes (if yes, please describe) 62.0% no

* Stickiness
* Hard to rake and compact
* Odor and pollution
* Failure due to mix and laydown temperatures

10. Have you experienced any problems in the compaction of mixtures containing rubber?

58.0% yes (if yes, please describe) 42.0% no

* Problems achieving density
* Roller pattern and equipment need modifications
* Mix has tendency to move in front of mat

11. Have you experienced any increase in pavement life using mixtures containing rubber?

18.0% yes (if yes, please describe) 82.0% no

12. Have you experienced any increase in cost using mixtures containing rubber?

100% yes (if yes, how much)

* Averages about twice as much as conventional

13. Did you have to do any modification to your equipment in order to mix, place, and compact the mixtures containing rubber?

62.0% yes (if yes, please describe) 38.0% no

* Pneumatic roller not allowed on mat
* Mixing rubber and asphalt required special equipment
* Asphalt plant pump had to be by-passed
* Separate storage of binder required

14. Are you familiar with the advantages associated with mixtures containing rubber? (if so please list)

* Resistance to aging and cracking
* Resistance to permanent deformation
* Increased elasticity

15. Approximately how many waste tires were generated in your state in the last year?

* It is estimated each state generates one tire per person each year (e.g., population of South Carolina: 3,512,000; therefore, SC generates approximately 3.5 million scrap tires per year).

16. If further research is conducted into the feasibility of the use of mixtures containing rubber, what specific characteristics would you like to see emphasized?

* Performance
* Cost effectiveness
* Impact on recycling
* Benefit/cost ratio
* Increased rutting resistance
* Plant and equipment modifications
* Differences among dry and wet processes
* Stripping problems
* Emissions control
* Service life
* Maintenance cost

LITERATURE REVIEW

Due to the restriction on the length of this article, only a portion of the extensive literature review conducted is presented in this section. Most of the discussed materials are covered in reference number 5.

Phoenix, Arizona has used asphalt rubber for many purposes, including pavement seal coats, SAM, SAMI, subgrade seals, lake liners, joint and crack sealants, roofing, and airport runway surface covers. There are several types of asphalt rubber used in Arizona. The following is a brief description of each process [6].

The McDonald process uses hot asphalt cement mixed with 25% ground tire rubber and diluted with kerosene to make the application easier. The Arizona refinery process uses a mixture of 18 to 22% ground rubber and hot asphalt cement and diluted with an extender oil. The asphalt rubber chip seal is the application of hot asphalt rubber followed by an application of hot precoated aggregate. SAM is a hot asphalt rubber chip seal applied to a stressed surface. SAMI is the same as a SAM, but it is followed by an asphaltic concrete overlay.

In Arizona, ten to twelve years of maintenance-free life can be expected from an asphalt rubber seal. Normal chip seals last six to eight years with some maintenance [6]. The initial cost of asphalt rubber is twice as much as conventional chip seal but will equal conventional costs over a 12 year period. Indications show that asphalt rubber life expectancy can go beyond 12 years by resealing it with standard or asphalt rubber seal [6].

Takallou and Hicks discussed a type of rubber-modified asphalt which is a mixture prepared by using 3 to 4 percent by weight relatively large rubber particles to replace some of the aggregate [7]. They indicated that the key to a successful rubber-modified asphalt is a low percentage of voids in the total mix (2 to a maximum of 4%). They also reported that the required level can be reached by increasing the mineral filler and the asphalt cement content, and by carefully following the compaction guidelines such as [7]: 1) avoiding the use of rubber tired rollers, 2). ensuring all water nozzles are working, 3) using liquid detergent in the drum water, and 4). Using a special wetting agent, Dweko, in the drum water.

In Nebraska, the Department of Roads is conducting a test on a section of Highway 4 using rubberized asphalt [8]. The main reason for experimenting with the rubber-asphalt mixture was to demonstrate the ability of these mixtures to retard reflective cracking compared to the conventional asphaltic concrete mixtures. For this project, 18% ground rubber and 82% liquid asphalt were blended together, heated at 204 °C (400 °F) for 45 minutes, and then mixed with the aggregate. They reported that resurfacing with rubber asphalt costs about three and a half times as much as resurfacing with conventional asphaltic concrete. At this time, the effects of the asphalt-rubber mixture on the performance of the pavement are not yet known [8].

McQuillen, et al., reported that the Alaska Department of Transportation and Public Facilities (ADOTPF) is evaluating rubber-modified hot-mix pavement as an alternative to conventional asphalt [9]. The main disadvantage in using rubber-modified mixes is its high initial cost [9]. They reported that in order to be cost effective, it must have a pavement life of 20 to 23 years, compared to 15 years for conventional mixes [9]. They also indicated that to make rubber-modified mixes more economically feasible, the lifecycle and/or capital costs must be reduced. The life-cycle costs could be reduced by including the intangible benefits, such as reduced traffic noise, improved skid resistance, reduction of sand, salting, and winter maintenance costs, and elimination of a solid waste material (i.e., rubber tires) [9].

The ADOTPF tested seven sections of roadway which were constructed between 1979 and 1981 and which included 3 to 4 percent coarse rubber particles [10]. It was hypothesized that this rubber-modified asphalt would increase skid resistance and durability as well as provide ice control [10]. Ice control was a primary concern for this project. Samples were loaded to failure at selected tensile strain levels and results showed the fatigue life to be more than 10 times greater than that of normal mixes. In addition, observations by the public indicated a reduction in skidding incidents and improved traction. Tests to measure stopping distances showed an average 25% reduction in stopping distances. However, there was a 50% increase in cost over conventional pavements [10].

Shuler, et al., evaluated six types of paving systems containing ground tire rubber [11]. These six systems included: a) asphalt-rubber seal coat, b) asphalt-rubber interlayer, c) asphalt-rubber concrete, d) asphalt-rubber friction course, e) asphalt concrete rubber filled, and f) friction course rubber filled. Systems a through d used 18-26% ground tire rubber, and systems e and f used rubber as an aggregate substitute [11].

Sixteen state highway departments pooled their funds for a research study administered by the FHWA. There were 219 test sections evaluated across the U.S. that were constructed between 1977 and 1984 [11]. For the interlayers, the performance of rubberized sections compared to control sections were not as good. This negative performance is believed to be due to inappropriate construction practice or intended use. Many of the interlayer test sections with negative results were constructed before 1979. This factor is important, because new technology developed after that date, and preblending became routine in the summer of 1979. Preblending improves the application process [11]. For asphalt-rubber seal coats, there was a slight negative performance, but it is believed to be due to construction practices. Flushing was the main cause of negative performance and it is hypothesized it was caused by inappropriate application quantities of binder and aggregate [11].

Vallerga, et al., evaluated several systems and their ability to reduce reflective cracking [12]. These systems included the SAM, the SAMI, the Plant-Mixed SAM, the SAM with open-grade, and plant-mixed surface. Conclusions made from field evaluations indicated that [12]: a) the SAM is very effective at reducing reflective cracking, as long as the existing cracks are no wider than 3/8 inches, b) the SAMI is very effective and practically eliminates all reflective cracking when used with the proper overlay and when wide cracks are filled in the existing pavement structure, c) the plant-mixed SAM has potential for reducing reflective cracking when it is used on a prepared surface.

Witczak's research results indicated that many factors affect the cost of using a pavement containing scrap rubber tires. These factors include time, location, contractor risk, project size, and project type [13].

A study by several Minnesota state agencies was conducted to investigate the use of recycled tire rubber in asphalt paving mixtures [14]. To yield a higher stability in the rubber mixtures, a coarser aggregate than normal was used. The larger stones were used to create a better aggregate interlock and the small rubber particles were supposed to fill in the gaps. It helped some, but the rubber mixtures still yielded very low stability values, and the surface texture was open and porous. The cost for the rubber mixtures was 35% to 50% higher than the control mix. The rubber mixtures did not perform well. They did not meet Minnesota DOT bituminous specifications. They exhibited low densities, low tensile strengths, and high air voids [14].

Wisconsin DOT performed a study of several sections of pavements containing RAP and rubberized mixtures [15]. The results indicated that the asphalt-rubber paving mixtures can be prepared and placed on the roadway. In addition, these projects demonstrated that scrap rubber tires can be used in a blend with asphalt cement; this blend may then be incorporated into a recycled asphalt paving mixture. No major problems were encountered during the paving operation. However, visible emission problems were noted at plants while producing the asphalt-rubber mix. On these projects, the engineers did not obtain asphalt-rubber paving mixes showing improved properties compared to the normal recycled asphaltic concrete mixtures [15].

Several state agencies in Connecticut and the Civil Engineering Department of the University of Connecticut performed research projects in the areas of: 1) overlays; 2) stress-relieving interlayers; 3) seal coats; and 4) joint seals [16-18]. The laydown cost for 1% rubber mixes was approximately 30% higher, and 60% higher for 2% rubber mixes compared to conventional mixtures. However, in general, the results were favorable despite the increase in cost.

The Missouri Highway and Transportation Department (MHTD) placed a test section of asphaltic concrete overlay modified with reclaimed ground rubber [19]. Performance indicators and measurements showed that both the control and the test section experienced slight rutting, probably occurring during the first week of traffic. The test section also showed some flushing in the wheelpaths, which gave the surface a slick appearance. Friction numbers for the test section were significantly lower than were those obtained from the control section, and some reflective cracking was observed in the test section. The construction costs of the asphalt-rubber test section were approximately 3.5 times as much as conventional asphaltic concrete mixtures. Much of the increase in cost was attributed to mobilization. Five special vehicles and trailers were brought in from International Surface, Inc., from Arizona [19].

In New York, two pilot test sections were constructed in order to help in the evaluation of the use of the new material. These test sections along with similar rubber-modified asphalt projects in other states were used in order to produce a basis on which to build cost evaluations. In review of these projects, it was established that lack of experience with the material and its behavior contributed to much of the inflation of cost. For this reason, realistic survey of increased costs based on the addition of ground tire rubber to asphalt concrete mixtures is not possible until construction of the material becomes a more familiar practice. The pilot projects reflected bid prices for the asphalt-rubber mixes to be 50 to 100 percent higher than conventional asphalt mixes; however, in-place costs for the rubber mixes were established as 25 to 50 percent higher than conventional hot mixes [20].

The Texas State Department of Highway and Public Transportation (TSDHPT) estimates that the number of scrap tires produced annually in Texas could be reduced by one-fifth if the highway department made it a practice to replace 10 percent of the paving asphalt cement with asphalt rubber [21-22]. The Texas SDHPT overall favors asphalt rubber interlayers to chip seals. Six of the 24 highway districts use the interlayers on a fairly regular basis, and the groups that have used the SAMIs believe that the material is working in delaying reflective cracks. Unlike the SAMs and SAMIs, asphalt-rubber crack sealant is used regularly by all 24 of the highway districts. It is preferred over all other products for sealing cracks in asphalt concrete and portland cement concrete pavements.

The California DOT uses its rubber-modified asphalt mixtures for overlays. From experimental evaluations, the Department has noted improvements in abrasion resistance and in lower permeability over conventional asphalt [23]. These results have also meant less maintenance work, and longer predicted pavement life due to decreases in water infiltration and oxidative aging. Experiments with thickness of asphalt-rubber overlays has also been a focus for the Department. Conclusions from experiments indicate that asphalt-rubber mixtures require less thickness than conventional asphalt mixtures; however, reduction percentages have not yet been established [23].

The Florida DOT contracted the National Center for Asphalt Technology (NCAT) to research ground tire rubber use in asphalt concrete mixtures [24]. Following this review in 1989, the Florida DOT organized two demonstration projects [25]. Conclusions from the first project were that the best additive for fine-grade surface friction courses of those tested was the minus 80 mesh ground tire rubber pre-blended at five percent (by weight of the asphalt cement), and that higher percentages influenced problems with construction. The results from the second project indicated that the minus 80 mesh blend at ten percent may have been the best candidate of those tested for open-graded friction courses. Construction also highlighted that large amounts of binder with increased amounts of ground tire rubber could cause an excess of asphalt which in turn could cause flushing in the wheelpaths [25].

In 1984, the Minnesota DOT tested asphalt-rubber as a binder in a dense-graded mix [26]. Both overlaid and reconstructed sections were used. No significant differences were noted for the overlays; however, less cracking was observed for the reconstructed pavements containing rubber in comparison to conventional pavement. Two other demonstration sections were constructed in which the finer components of dense graded aggregate were replaced by ground tire rubber before asphalt was added (PlusRide system). No substantial improvements were observed by field personnel [26].

The Federal Highway Administration published a report indicating the technical guidance to asphalt pavers on how to use "Crumb Rubber Additive" (CRA) technology [27]. Beside the dry and wet process there are newer applications in the use of rubber in asphaltic concrete mixtures and they are generally referred to as the "Generic Dry Process," "Chunk Rubber Asphalt Concrete", and "Continuous Blended Asphalt Rubber" [27].

SUMMARY, CONCLUSIONS, AND RECOMMENDATIONS

Based on the findings of the survey and literature review, the SCDHPT initiated a demonstration project which included paving a test section (approximately 0.9 miles) using the dry process (PlusRide process). The paving was completed by May 4, 1992; however, there is not much technical information to report at this time.

1. The following is a summary of the responses given to a questionnaire mailed to all state and federal agencies

involved with highway construction:

a) Fourteen (14%) percent of respondees indicated that they are using mixtures containing rubber for flexible pavements.

b) The disadvantages of using this type of mixture, mentioned by the respondees, included cold temperature cracking, shorter compaction period, constructibility and increased mix temperature.

c) The problems encountered by the agencies using mixtures containing rubber included cost, increased temperature and viscosities, excessive plant emissions, and limited supply of rubber.

d) The problems during placing the mixtures encountered by the agencies included the compaction difficulties, stickiness, odor and pollution.

e) All of the respondees indicated that the increase in cost is a major concern of their agencies.

f) The future research topics indicted by the respondees included studies of rutting resistance of the mixtures, cost effectiveness, emissions control, recyclability of the asphalt-rubber mixtures, service life, stripping problems, and performance.

2. Based on the limited trials conducted around the nation, in general, the performance of flexible pavements containing scrap tires has been satisfactory. However, there were some cases in which engineers concluded that the performance was not acceptable.

3. There are two processes involved with the use of tires in asphaltic concrete mixtures referred to as dry and wet process. In the dry process, the rubber is added to the aggregate and in the wet process, the rubber is added to the asphalt cement. There are some experimental pavements containing these two processes around the nation. In general, both of the processes proved to be effective.

The dry process uses many more tires than the wet process. For instance, for a typical paving job requiring 5,000 tons of hot mix, if the dry process is being used (3% rubber by total weight of the mixture), approximately 25,000 tires could be utilized. However, for the same mixture, if the wet process is chosen, then approximately 8,000 tires could be used (based on 6% asphalt cement content and 16% rubber by total weight of asphalt cement).

4. The cost of either process is much higher than the conventional paving process. In some cases, the cost of using the rubber in asphaltic concrete mixtures doubled the cost of

the construction. However, it is anticipated that the cost would be higher by 20-40% in the future due to wider use of these materials in pavements. The reasons for the higher costs include increased volume of asphalt cement, addition of rubber, increased energy to heat to higher temperatures and extended mixing time, increased plant labor/ equipment to handle, and increased royalties (patented systems only).

5. Some of the <u>reported</u> potential benefits of using the mixtures containing rubber include conservation of asphalt cement and aggregates; preservation of the environment; thinner lift; increased pavement life; retarded reflection cracking; decreased traffic noise; reduced maintenance costs; and decreased pollution.

6. Some of the disadvantages of using such a system include the initial high cost; expertise of contractors; necessary modifications to equipment or plants; lack of specifications by agencies such as ASTM; and potential problems with recyclability of these materials.

7. For mixtures containing rubber to be cost effective, either the design life must exceed that of an equivalent alternative or the layer equivalency of this type of a mix to a conventional application must be significantly greater than a value of unity.

8. Research to develop new testing procedures is necessary, as well as an accurate method for extracting and evaluating field samples. Currently, the processes using rubber in asphalt rely heavily on manual labor. This can result in several problems during the production process. Automation would keep these problems to a minimum. Compaction of rubber mixes is a problem because of the stickiness. Rubber tired rollers are not recommended, and steel wheel rollers should have plenty of water available to the wet drum, as well as a mild detergent and/or a wetting agent.

9. It is not known if rubberized asphalt can be recycled in the future. Further research is needed in this area. Lastly, environmental and health studies need to be performed to determine if the emission production falls within state/federal environmental and health limits. There have been several reports of increased air pollution and visible emissions at the plant site.

ACKNOWLEDGEMENTS

The author wishes to extend his sincere appreciation to the South Carolina Department of Highways and Public Transportation (SCDHPT) officials and Michelin Tire Corporation of Greenville, South Carolina, for sponsoring and funding this research project. The author also wishes to thank Ms. Catherine Nisbet, Mr. Robert Wentz, and Mr. Stephen Roberts of Clemson University for their assistance and cooperation during the project. The assistance of those state and federal agencies that participated in the questionnaire is greatly appreciated. The assistance of Mr. Stewart, Mr. Gibson, Mr. Fletcher and Mrs. Gambill of the SCDHPT is also appreciated.

REFERENCES

[1]. Focus, Strategic Highway Research Program, July 1991, Washington, D.C.

[2]. "Rubber Roads Debated," Engineering News Record, August 5, 1991, p.10.

[3]. "Use, Availability and Cost Effectiveness of Asphalt Rubber in Texas," C.K. Estakhri, E.G. Fernando, J.W. Button, and G.R. Teetes, Texas Transportation Institute, Research Report 1902-1F, Texas A&M University, College Station, Texas, 1990.

[4]. "State of the Practice - Design and Construction of Asphalt Paving Materials with Crumb Rubber Modifier," M.A. Heitzman, U.S. Department of Transportation, Federal Highway Administration, Publication No. FHWA-SA-92-022, May 1992.

[5]. "A Feasibility Study of the Use of Waste Tires in Asphaltic Concrete Mixtures," S.N. Amirkhanian, Report No. FHWA-SC-92-04, May 1992.

[6]. Russell H. Schnormeier, "Fifteen-Year Pavement Condition History of Asphalt-Rubber Membranes in Phoenix, Arizona", Transportation Research Record 1096, 1986, p.62.

[7]. H.B. Takallou and R.G. Hicks, "Development of Improved Mix and Construction Guidelines for Rubber-Modified Asphalt Pavements", Transportation Research Record 1171, 1988, p.113.

[8]. "New Life For Old Tires", AASHTO Quarterly, January 1991, p.24.

[9]. Jay L. McQuillen Jr., H.B. Takallou, R.G. Hicks, and Dave Esch, "Economic Analysis of Rubber-Modified Asphalt Mixes", Journal of Transportation Engineering, Vol. 114, 1988, p. 259.

[10]. David C. Esch, "Construction and Benefits of Rubber-Modified Asphalt Pavements", Transportation Research Record 860, 1982, p.5.

[11]. T.S. Shuler, R.D. Pavlovich, and J.A. Epps, "Field Performance of Rubber-Modified Asphalt Paving Materials", Transportation Research Record 1034, 1985, p.96.

[12]. B.A. Vallerga, G.R. Morris, J.E. Huffman, and B.J. Huff, "Applicability of Asphalt-Rubber Membranes in Reducing Reflection Cracking", AAPT Vol.49, , 1980, p.330.

[13]. M.W. Witczak, "State of the Art Synthesis Report Use of Ground Rubber in Hot Mix Asphalt", Department of Civil Engineering, University of Maryland, June 1, 1991.

[14]. C.M. Turgeon, "Waste Tire and Shingle Scrap/Bituminous Paving Test Sections on the Willard Munger Recreational Trail Gateway Segment", Minnesota Department of Transportation, February 1991.

[15]. Clinton E. Solberg and David L. Lyford, "Recycling with Asphalt-Rubber, Wisconsin Experience," Wisconsin Department of Transportation's Internal Report.

[16]. Jack E. Stephens, "Recycled Rubber in Roads - Final Report", Report #CE 81-138, Civil Engineering Dept., University of Connecticut, April 1981.

[17]. Jack E. Stephens and S. Mokrzewski, "The Effect of Reclaimed Rubber on Bituminous Paving Mixtures," Report #CE 74-75, Civil Engineering Dept., University of Connecticut, Feb. 1974.

[18]. Standard Specifications for Roads, Bridges and Incidental Construction, Form 811, State of Connecticut, Department of Transportation, 1974.

[19]. Mark Webb, "Asphalt Rubber Concrete Test Section," Experimental Project No. M090-01, Missouri Highway and Transportation Department, Interim Report, Project IR-70-3(146), March 1991.

[20]. "Waste Tire in New York State: Alternatives to Disposal", New York State Department of Environmental Conservation: Division of Solid Waste: Bureau of Resource Recovery, Albany, N.Y., 1987.

[21]. Shuler, T. S., Pavlovich, R. D., Epps, J. A., and Adams C. K. "Investigation of Materials and Structural Properties of Asphalt-Rubber Paving Mixtures," Research Report RF 4811-1F, Texas Transportation Institute, Texas A&M University, College Station, Texas, September, 1985.

[22]. Schuler, T. S., Gallaway, B. M. and Epps, J. A., "Evaluation of Asphalt-Rubber Membrane Field Performance," Research Report 287-2, Texas Transportation Institute, Texas A&M University, College Station, Texas, May, 1982.

[23]. Van Kirk, J. L., "CALTRANS Experience With Asphalt-Rubber Concrete - An Overview and Future Direction," Proceedings, National Seminar on Asphalt Rubber, Kansas City, Missouri, 1989, pp. 417-431.

[24]. Roberts, F. L., Kandhal, P. S., Brown, E. R., and Dunning, R. L., "Investigation and Evaluation of Ground Tire Rubber in Hot-Mix Asphalt," National Center for Asphalt Technology, Auburn University, Alabama, 1989.

[25]. Page, G. C., "Florida's Experience Utilizing Ground Tire Rubber in Asphalt Concrete Mixtures," Proceedings, National Seminar on Asphalt Rubber, Kansas City, Missouri, 1989, pp. 499-535.

[26]. Turgeon, C. M., "The Use of Asphalt-Rubber Products in Minnesota," Proceedings, National Seminar on Asphalt Rubber, Kansas City, Missouri, 1989, pp. 311-327.

[27]. "State of the Practice For the Design & Construction of Asphalt Paving Materials with Crumb Rubber Additive", US Dept. of Transportation, FHWA, July 1991.

Fouad M. Bayomy[1], Glenn D. Carraux[2]

MODIFICATION OF HOT MIX ASPHALT CONCRETE USING AN ETHYLENE BASED COPOLYMER

REFERENCE: Bayomy, F. M. and Carraux, G. D., **"Modification of Hot-Mix Asphalt Concrete using an Ethylene-Based Copolymer,"** Use of Waste Materials in Hot-Mix Asphalt, ASTM STP 1193, H. Fred Waller, Ed., American Society for Testing and Materials, Philadelphia, 1993.

ABSTRACT: Research was conducted to determine the feasibility and effect of using an ethylene based copolymer (EBC) derived from reprocessed low density polyethylene (LDPE) as an additive to improve asphalt mixture performance. The decision to use reprocessed LDPE to improve an asphalt pavement stems from both the deteriorated condition of the nation's highways, and the increasing need to develop alternative uses for materials now heading to the nation's overcrowded landfills.

Research included the determination of the method for applying the modifier to the hot mix asphalt concrete and its treatment levels. Primary concern, however, was the evaluation of the performance of the modified hot mix asphalt mixtures.

KEYWORDS: Modified hot-mix asphalt, ethylene based copolymer, reprocessed polyethylene

---

Previous research indicated that polyethylene copolymers as neat asphalt additives improved pavement performance [1,2,3]. Under this research program, testing was completed to determine a mixing method when using an ethylene based copolymer (EBC), an optimum treatment level and the resulting performance characteristics. Results showed that the EBC modifier should be mixed with the aggregate prior to mixing with asphalt to allow full dispersion of the modifier over the aggregate particles. The optimum treatment level is approximately 1% of the dry aggregate weight. Under these conditions, EBC modified mixtures showed significant improvement in mix characteristics in rutting resistance, fatigue cracking and age hardening. The use of the EBC modifier would satisfy the need to improve conventional asphalt mixtures, and the source of the modifier from reprocessed LDPE can benefit the current municipal solid waste problem.

---

[1]Associate Professor, Department of Civil Engineering, University of Idaho, Moscow, ID 83843

[2]Director of Engineering, Carraux Enterprise International, Inc., Houston, TX 77064

INTRODUCTION

In 1990, the road construction industry placed 500 million tons of asphalt pavements for both new construction and repair of existing road structures. This paving work, however, did not fulfill the annual roadway needs of this country. The Federal Highway Administration, which now spends $13 billion per year on repairs and construction, estimates that bringing the nation's roads up to minimum engineering standards will cost from $565 billion to $655 billion over the next twenty years. It is estimated that by 1993, more than 90% of the nation's roadways will have reached their design life, requiring repair, rehabilitation, or reconstruction. This volume of expenditures has both federal and state agencies looking at improvements in the road construction industry, including improvements in mix design calculations such as large stone mixes, the use of asphalt modifiers, and improvements in construction equipment.

Just as the nation's highways are in an ever increasing need of repair, the United States is faced with another critical infrastructure problem. More stringent regulations and limited capacity are closing landfills, this nation's primary solution to municipal solid waste. According to the Environmental Protection Agency, in five years, only one-fourth or 5000 landfills will remain of those in operation during the late 1970's. This situation requires the reevaluation of those materials currently discarded, for potential new uses.

Debate has continued over the use of waste materials in asphalt pavements. The use of reprocessed LDPE in the form of an EBC modifier should provide beneficial results to the performance of an asphalt pavement, and should not be merely a disposal alternative for this LDPE.

The objective of the research conducted was to investigate the use of an ethylene based copolymer additive derived from reprocessed low density polyethylene to improve pavement performance. The EBC material was developed in France under the name "Starflex" [1], and is currently in use in Western Europe [1,2,3]. To determine the feasibility of its use in the United States, this modifier was evaluated under accepted standards and tests for this country.

APPROACH

Figure 1 illustrates the major phases of the laboratory study undertaken in this program.

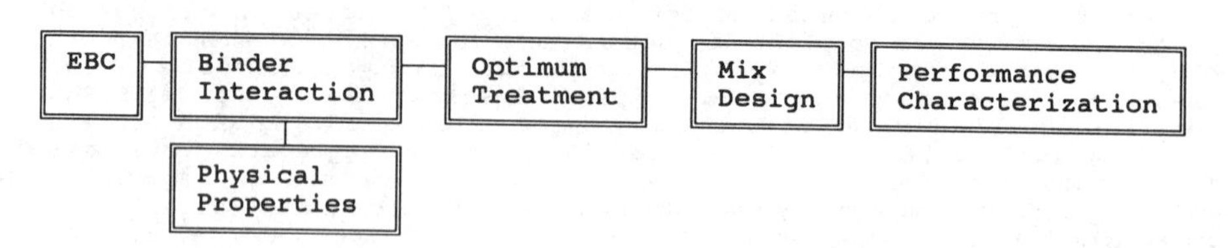

FIG. 1 -- Graphical Illustration of Laboratory Study

Interaction with Neat Asphalt

Because of the paraffinic nature of the EBC modifier, the interaction of the modifier with a neat asphalt will depend mainly on the type and properties of the asphalt. Accordingly, three different sources of asphalt cement were selected to cover a wide range of asphalt cements. Blends of the EBC modifier with each asphalt source and with three different grades (AC-5, AC-10 and AC-20) were prepared. The AC-20 grade was included as the control binder, normally used in bituminous mixtures in Texas. Based on compatibility tests, it was concluded that the modifier blends well with all sources of asphalt at low treatment levels. At high treatment levels, the blends tend to be thick and impractical for application, especially for high asphalt grades. As such, the EBC modifier should not be added to the asphalt like other polymer-based modifiers.

Optimum Treatment Level

In determining an optimum treatment level, investigative work was undertaken using a series of physical tests reported to correlate to roadway performance [4,5]. Additionally, based on previous research conducted in Europe at both the Centre de Recherches Routieres, Belgium [2], and the LAVOC Highway Laboratory of the Federal Polytechnic, Switzerland [3], it was determined that a 1% EBC modifier by weight of aggregate would improve asphalt mixtures. Accordingly, testing was completed using a control mixture, and mixtures modified with the EBC at 1, 2, 3, 4 and 5% by weight of the aggregate, added to a surface course mix with 5.5% (AC-10) asphalt and aggregate gradation following Type D, Item 340 of the Texas Department of Transportation (TxDOT) specifications [6]. Results of air voids, Hveem stability, modulus of resilience and indirect tensile strength for the control and modified mixes are shown in Figure 2. Results indicated an optimum treatment level of 1 to 3%, since for treatment levels of 4 and 5% air voids were relatively high. Laboratory samples also showed agglomeration and very rough surfaces.

Mix Design

To determine the mix design characteristics of the EBC modified mixes, three mix designs were developed using the Texas standard design procedures (Tex-204-F "Design of Bituminous Mixtures") [7]. These mixes were also used for performance evaluation of EBC modified mixes. Based on the investigative study results and the European research [2,3], two EBC treatment levels were selected, 1% and 3% by aggregate weight as being the practical range of the modifier application. A control mix following Type D, Item 340 of the TxDOT specifications [6] was developed. For this purpose, one asphalt type was selected and only one asphalt grade was used (AC-20) to represent a typical mix mostly used for an asphalt concrete surface course. The modified mixes were prepared by blending the EBC modifier with the hot aggregates (identical to those of the control mix) prior to mixing with asphalt. Aggregates weight adjustment was made to account for the modifier weight, hence the total aggregate weight in each prepared sample was the same for the three mixes.

The optimum binder content was determined using laboratory density, Hveem stability and resilient modulus test results. Optimum values are listed in Table 1. The TxDOT design procedures specify that the optimum binder content corresponds to 96% density (or 4% air voids) provided minimum Hveem stability criteria is satisfied. Since Hveem stability is primarily a measure of aggregate internal friction, these tests were supplemented with resilient modulus test according to the ASTM Method for Indirect Tension Test for Resilient Modulus of Bituminous Mixtures (D 4123). The resilient modulus test results revealed significant increases can be expected with the addition of the EBC modifier.

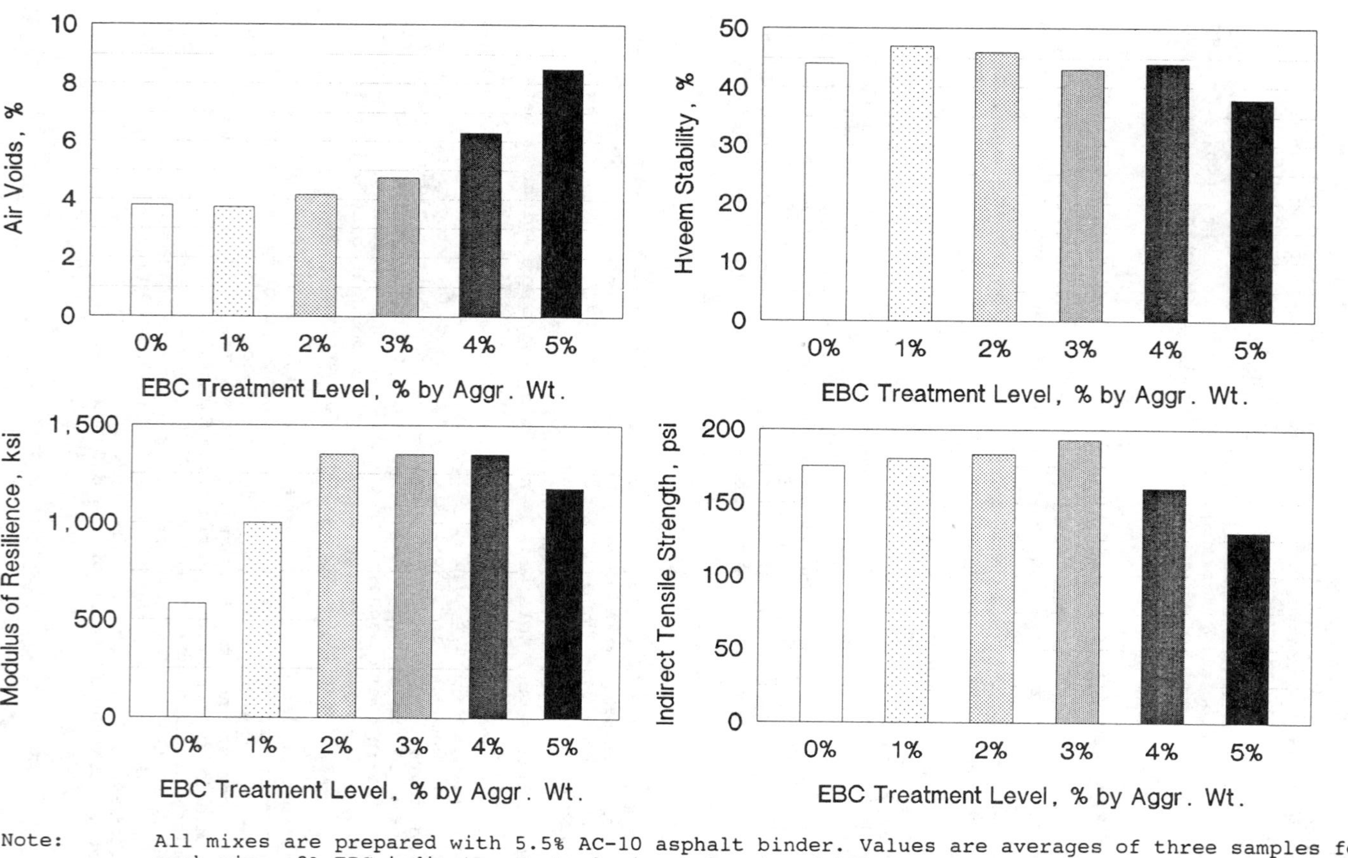

Note: All mixes are prepared with 5.5% AC-10 asphalt binder. Values are averages of three samples for each mix. 0% EBC indicates Control mix. (1 psi = 6.89476 KPa, 1 ksi = 6.89476 MPa)

FIG. 2 -- Effect of EBC Treatment Level on Mix Properties

TABLE 1 -- Mix Design for EBC Modified Mixtures

Property	Control	1% EBC	3% EBC
Opt. AC% by Aggr. Weight (AC-20)	5.4	5.1	5.5
Hveem Stability	40	51	41
Blk. Sp. Gravity	2.350	2.300	2.220
Air Voids %	4	4	4
Mr, ksi *	820	1220	2060

* 1 ksi = 6.89476 MPa

PERFORMANCE CHARACTERISTICS

Having determined an optimum treatment level for the modifier and optimum binder level for a modified mixture, testing was completed to determine the performance characteristics of the modified mixtures.

Performance evaluation was conducted on mixes using 1% and 3% EBC modifier by aggregate weight. These mixes were compared to a control mix. Mix design parameters of the selected mixes were as given in Table 1. Each mix performance was evaluated using Modulus of Resilience (Mr), Indirect Tensile Strength (ITS), Dynamic Creep (Rutting), Fatigue and Aging Tests. These tests would be the key to determining the effectiveness of an EBC modifier in improving performance.

Modulus of Resilience

This property measures the ability of the asphalt mix to withstand the repetition of traffic loads. An increase in stiffness implies that the mixtures' resistance to permanent deformation has been improved. To evaluate the resilient modulus of the designated mixtures, the test was conducted on mixtures at 77°F and 104°F, and performed in accordance with ASTM D 4123 test method. As shown in Table 2, the results indicated that the addition of an EBC modifier improved the mixtures' temperature susceptibility and provided better resistance to permanent deformation.

TABLE 2 -- Summary of Mr and ITS Tests

Property	Test Temp.	Control Mix	1% EBC	3% EBC
Modulus of Resilience Mr (ksi) *	77°F (25°C) 104°F(40°C)	1077 147	2032 723	2304 919
Indirect Tensile Strength (psi) **	77°F (25°C) 104°F(40°C)	163 41	174 63	234 137

* 1 ksi = 6.89476 MPa
** 1 psi = 6.89476 KPa

Indirect Tensile Strength (ITS)

Indirect Tensile Strength tests were performed by loading the samples diametraly at a rate of 2 in./min. (5.08 mm/min) until failure in accordance with the loading procedures of the ASTM Test Method for the Affect of Moisture on Asphalt Concrete Paving Mixtures (D 4867). Generally, ITS is expected to decrease with temperature increase. The rate of decrease would be a function of the mix property. Results of ITS (Table 2) showed similar trends to those obtained with Mr results. This confirmed that mixes with an EBC modifier would be stiffer and have higher strength than the control mix.

Permanent Deformation

The potential of a mixture to resist permanent deformation is a function of many factors. These factors are related to mix design properties and to the existing operating conditions such as temperature, traffic load variation, moisture change and asphalt oxidation. In this program, mixes were subjected to dynamic creep (rutting) testing using the indirect tension diametral dynamic loading test [8]. Rutting process in asphalt mixtures develops in three stages. During the first stage, which is referred to as "primary", rutting occurs, mainly, due to densification and particle re-orientation. In this stage permanent deformation progresses rapidly until mixtures reaches a steady state. The steady state rutting (second stage) develops at much slower rate and occurs along long period. The rutting in the second stage continues until reaches a critical value after which the mix cracks. If the load continues after the mix has had a critical permanent deformation, cracks would propagate causing fatigue cracking failure. This is the third stage of rutting process where failure is more due to crack propagation in addition to the excessive deformation that has developed in the mix. A quality of a mix to resist rutting can be characterized by the number of cycles at the end of the steady state stage (N_s) and the amount of deformation that occurred at that point (D_p). Since these two parameters are function of the load applied, testing temperature and the geometry of the test, a dimensionless factor has been developed to reflect the effectiveness of the EBC modifier on the rutting resistance of the mixtures. This dimensionless factor is defined as: "Relative Rutting Potential (RRP) is the ratio of the permanent deformation at the end of steady state process (Dp) of the modified mix to that of an identical control mix under the same testing conditions", that is:

$$RRP\ (\%) = (D_{pm} / D_{pc}) \times 100$$

where: D_{pm}: permanent deformation of the modified mix determined at its N_s cycles, and

D_{pc}: permanent deformation of the control mix determined at its N_s cycles.

To compare the three mixes, one test load was selected to test the three designated mixes at two temperatures. Results as shown in Table 3 illustrate the potential of rutting resistance of the modified mixes as compared to the control unmodified mix. This means that if the rutting expected in a control mix is a 100%, the rutting potential in the 1% EBC modified mix would be between 63% to 73% of that of the control mix depending upon the temperature. The rutting potential, however, would be reduced to about 54% to 42% if a 3% EBC modifier is used. Results also indicate that not only the amount of rutting would be reduced, but also the life of the mix would be significantly increased. It is noticed that while 2,000 cycles was sufficient to reach critical rut depth at 77 °F in the given test, the 1% EBC and 3% EBC modified mixes lasted up to 32,000 and 550,000 cycles respectively. At higher temperature the 3% EBC modified mix was more effective. This indicates less temperature susceptibility and higher resistance to permanent deformation of the EBC modified mixes.

Table 3 -- Relative Rutting Potential (RRP) of the EBC Modified Mixes

Mix	Test Temp. 77°F (25°C)			Test Temp. 104°F (40°C)		
	N_s*	D_p(in)**	RRP (%)	N_s*	D_p(in)**	RRP (%)
Control	2,000	0.092	100	600	0.0825	100
1% EBC	32,000	0.058	63***	673	0.060	73
3% EBC	550,000	0.050	54	250,000	0.035	42

* Cycles
** 1 in = 25.4 mm
*** RRP for 1% EBC mix at 77 °F = (0.058/0.092) x 100 = 63 %

Fatigue

The repetition of an induced elastic strain in an asphalt pavement will lead to fatigue failure. Depending on the value of the induced tensile strain and the frequency of load repetition, a fatigue life of a certain mixture can be evaluated. Generally, load frequency is a traffic-related parameter, while induced tensile strain depends on the applied stress level, mix type, pavement temperature and the structural design of the pavement itself.

The indirect tensile repeated load test was utilized to produce failure in disc samples [9]. Similar to the Mr test except that the load is continued until sample failure, the three designated mixtures were tested at three load levels based on the sample's indirect tensile strength. Results in Table 4 indicate that the addition of an EBC modifier extended the mixtures' fatigue life.

TABLE 4 -- Fatigue Test Results for Control and EBC Mixes

Mix	Stress Level 1		Stress Level 2		Stress Level 3	
	psi*	N_f**	psi*	N_f**	psi*	N_f**
Control	8	53184	12	59872	40	2199
	8	35359	12	50120	40	5400
	8	44252	12	22501	40	11780
1% EBC	24	29475	40	18652	60	4127
	24	39842	40	67885	60	7237
	24	73720	40	63720	60	6742
3% EBC	24	121165	40	253668	60	16862
	24	32621	40	182487	60	20226
	24	338550	40	295533	60	14815

* 1 psi = 6.89476 KPa
** Cycles

Aging Effect

Temperature effect on an asphalt mixture has two holds. First, temperature causes softening of the asphalt and reduces asphalt viscosity. Second, asphalt oxidizes in a hot air environment, causing mixtures to become more brittle as they age. To separate these two effects, samples were aged under the same conditions, then tested at different temperatures.

The aging effect was performed through the evaluation of Mr for both aged and unaged samples. Results of Mr values for the designated mixtures before and after aging are presented in Table 5. An aging factor defined as the ratio of Mr value after aging over that value before aging was calculated for all test conditions. The aging test results revealed that EBC modified mixes were less susceptible to aging than the control mix.

TABLE 5 -- Summary of Aging Test Results

Mix	Test Temp. °F (°C)	Mr (ksi)* Before Aging	Mr (ksi)* After Aging	Aging Factor
Control	77 (25)	355	1203	3.39
1% EBC	77 (25)	697	1358	1.95
3% EBC	77 (25)	1338	1708	1.28
Control	39 (4)	3348	3221	0.96
1% EBC	39 (4)	3773	3181	0.84
3% EBC	39 (4)	2433	2203	0.91

* 1 ksi = 6.89476 MPa

PERFORMANCE PREDICTION

To determine the feasibility of using an EBC modifier, it is not only important to determine the performance of the mix samples according to accepted test methods, but it is necessary to predict the performance of a pavement using the modifier. A hypothetical pavement design using a control mixture and an EBC modified mixture was developed. These designs were then evaluated to determine the effect on the pavement's service life.

Basically, the approach adopted is to assume traffic and soil conditions as well as base and subbase conditions. Using the AASHTO Design Guide [10] and the Asphalt Institute Design Manual [11], three design alternatives were developed. An evaluation of each design is performed using the AASHTO limiting strain criteria [12] to determine fatigue and rutting life for each design. In order to determine the stress-stain state in the developed pavement sections, the computer program CHEVPC was used. CHEVPC is a PC version adapted from the original CHVRON multilayer elastic analysis FORTRAN computer program [13,14]. Each developed system was input and the strains at certain locations were determined. For each system, the tensile elastic strain at the bottom of the AC layer and the vertical compressive strain at the top of the subgrade were determined. Based on the resulting strains, the fatigue and rutting lives were determined.

Table 6 presents the results. The service life of the EBC modified mixes are much greater than those of the control mix. The gain in service life is a very good indicator of the economic benefit of an EBC modified mixes. Simply, it will lead to significant reductions in maintenance costs.

TABLE 6 -- Performance of EBC Mixes in Different Design Alternatives

Design Alterna -tive	HMA Surface Layer	Layer Modulus (ksi)* (77°F) (25°C)	Rutting Life		Fatigue Life	
			Compr. Strain, micros	N_r, EAL_{18} (MM)**	Tensile Strain, micros	N_f, EAL_{18} (MM)**
Design Alt. No.1	Control	1070	458	0.18	190	0.4
	1% EBC	2030	378	1.0	130	1.0
	3% EBC	2304	362	1.2	120	1.5
Design Alt. No.2	Control	1070	189	20	71	60
	1% EBC	2032	115	60	41	100
	3% EBC	2304	105	80	36	100
Design Alt. No.3	Control	1070	410	0.4	137	0.4
	1% EBC	2030	193	9	88	4
	3% EBC	2304	115	60	80	8

* 1 ksi = 6.89476 MPa
** MM = Millions

ECONOMIC CONSIDERATIONS

Much debate has occurred regarding the use of waste materials in hot mix asphalt. From glass to rubber to plastics, numerous materials have been proposed for use. Two of the aspects of this debate include: 1) that any additive proposed for hot mix asphalt should not sacrifice the performance of that pavement, and 2) that the additive not unnecessarily increase the price or place hot mix asphalt in a disadvantaged price position. The solution for solid waste problems should not be at the expense of the hot mix industry, solving one problem while possibly creating another [15].

Tests on the EBC modifier satisfy these objectives. Based on current accepted test methods, the use of an EBC modifier can provide improvements to the performance characteristics of a pavement mixture. Although there is an increase in the initial cost of construction, estimated as an addition of approximately 30% based on European construction projects and estimated for U.S. construction, the economic benefits lie in the increased service life of the pavement, decreased maintenance costs, and decrease in such items as user delay costs, the cost which the traveling public incurs when delayed by rehabilitation or reconstruction activities.

CONCLUSIONS

The purpose of the research on an EBC modifier was two-fold. First, the testing was completed to determine the improvements which could result from the use of an ethylene based copolymer in an asphalt pavement mix. Secondly, the EBC modifier tested was formulated from reprocessed LDPE to determine the effect, if any, of using a waste material in asphalt modification.

The use of an EBC can improve a number of performance properties of asphalt mixtures. As a modifier added to the aggregate, the EBC alters the mix design significantly. Its use reduces the binder demand in the vicinity of 0.5% by weight of the neat asphalt, with an optimum treatment level for the EBC of approximately 1% by weight of the aggregate. The EBC modifier improved the mixture's temperature susceptibility and provided better resistance to permanent deformation. Modified mixtures show an extended fatigue life, and were less susceptible to age hardening.

The combination of results provides for improved performance of a hot mix asphalt pavement.

The formulation of the EBC modifier provides for the use of reprocessed low density polyethylene. In 1988, more than 7.1 billion pounds, or 93% of the LDPE produced in the United States, was disposed of in the municipal solid waste stream.

The United States is facing a crisis in its infrastructure, in both the highway sector as well as in the municipal solid waste disposal industry. Seemingly unrelated, they share a common tie. The EBC modifier derived from reprocessed low density polyethylene can benefit both sectors by improving the nation's highway system while reducing the demand on the nation's overcrowded landfills.

ACKNOWLEDGEMENT

Research for this paper conducted at the Center for Construction Materials Technology, Southwestern Laboratories, Inc., Houston, Texas. Research sponsored by Carraux Enterprise International, Inc., Houston, Texas.

REFERENCES

[1] "Starflex", Northstar Civil Engineering, Ltd., London, United Kingdom.

[2] Sterrebeek, "Test Report E 2125", Centre de Recherches Routieres, Belgium, March, 1987.

[3] Pigois, "Study of the Behaviour of Bituminous Concrete Produced with B 80/100 Bitumen to which Starflex Polymers have been added", LAVOC Highway Laboratory, Department of Civil Engineering, Federal Polytechnic, Lausanne, Switzerland, January, 1988.

[4] Martinez, D.F. and Tahmoressi, M., "Test Reported to Correlate to Asphalt Performance", Strategic Highway research Program A-004, January, 1990.

[5] ARE Inc., Engineering Consultants, "Asphalt properties and Relationship to Pavement Performance Literature Review", Report TR-ARE-A-003A-89-3, Strategic Highway Research Program, August, 1989.

[6] ______, "Standard Specifications for Construction of Highways, Streets and Bridges", Texas State Department of Highways and Public Transportation (SDHPT), Austin, Texas 1982.

[7] ______, "Manual of Testing Procedures, Volume 2", Texas State Department of Highways and Public transportation (SDHPT), Austin, Texas 1982.

[8] Jallejo, J., Kennedy, T.W., "Pavements Deformation Characteristics of Asphalt Mixtures by Repeated Load Indirect Tension Test", Report 183-7, Center for Highway Research, The University of Texas at Austin, Austin, Texas, June 1976.

[9] Rauhut, J.B. and Kennedy, T.W., "Characterizing fatigue Life for Asphalt Concrete Pavement", Transportation Research Record 888, Transportation Research Board, Washington, D.C., 1982.

[10] ______, "AASHTO Guide for Design of Pavement Structures 1986", American Association of State Highway and Transportation Officials, Washington, D.C., 1986.

[11] ______, "Thickness Design - Asphalt Pavements for Highways and Streets", Manual Series No. 1 (MS-1), September 1981 edition, The Asphalt Institute, Lexington, Kentucky.

[12] Havens, J.H., Dean, R.C. and Southgate, H.F., "Pavement Design Schema", Structural Design of Asphalt Concrete Pavement to Prevent Fatigue Cracking, Highway Research Board, Special Report 140, 1973.

[13] Michelow, J., "Analysis of Stress and Displacements in an n-Layered Elastic System Under a Load Uniformly Distributed on a Circular Area", California Research Corporation, Richmond, California, 1963.

[14] Painter, L.J., "CHEVRON N-Layer Program - Improved Accuracy", California Research Corporation, Richmond, California, 1980.

[15] Warren, J., " The Use of Waste Materials in Hot Mix Asphalt", Special Report 152, National Asphalt Pavement Association, Lanham, Maryland, June, 1991.

James R. Lundy[1], R. G. Hicks[1], H. Zhou[2]

GROUND RUBBER TIRES IN ASPHALT-CONCRETE MIXTURES - THREE CASE HISTORIES

REFERENCE: Lundy, J. R., Hicks, R. G., and Zhou, H., **"Ground Rubber Tires in Asphalt-Concrete Mixtures - Three Case HIstories,"** Use of Waste Materials in Hot-Mix Asphalt, ASTM STP 1193, H. Fred Waller, Ed., American Society for Testing and Materials, Philadelphia, 1993.

ABSTRACT: Results are presented from three case histories in which ground rubber modification was used in the construction of asphalt concrete pavements. Material properties, performance, benefits, limitations, and costs are discussed for each project. The mixtures used in these projects cover crumb rubber modification of gap-graded aggregates, rubber modified binders, and the so-called generic crumb rubber mixtures. With some noteworthy exceptions, all the mixtures are performing well.

KEYWORDS: Crumb rubber, asphalt concrete, diametral modulus, fatigue, rubber-modified binder.

The last ten years has seen significant changes in rubber-modified binders and asphalt mixtures [1]. Suppliers of these products have changed binders and the specification of mixtures in response to the perceived needs of user agencies. The extent to which these needs have been met can only be demonstrated through continued monitoring of the performance of field trials.

In the Pacific Northwest, agencies have experimented with both the wet process (crumb rubber is incorporated into the asphalt cement prior to mixing with the aggregate) and the dry process (crumb rubber is added to the heated aggregate prior to being mixed with a conventional binder). City, county, state and federal agencies have all successfully placed asphalt paving materials with the crumb rubber modifier. The purpose of this paper is to describe the performance of three projects which were constructed over the last ten years in the Pacific Northwest. They encompass some of the diversity present in rubber-modified asphalt mixtures. Projects have been constructed using 1) ground rubber tires

[1] Assistant Professor and Professor, respectively, Civil Engineering Department, Oregon State University, Corvallis, OR 97331

[2] Senior Engineer, Nichols Consulting Engineers, Reno, NV 89509.

and tire buffings to replace aggregate in a gap-graded gradation, 2) a modified binder consisting of fine rubber tire buffings mixed with ordinary asphalt cement, and 3) crumb tire rubber added to a dense aggregate gradation. The findings from the case studies are used to develop improved guidelines for the use of ground rubber tires in asphalt-concrete mixtures.

MT. ST. HELENS PROJECT

This project was constructed as part of the Volcanic Activity Disaster Relief (VADR) projects in 1983 following the eruption of Mt. St. Helens. The project consists of 1.8 kilometers of crumb rubber modified asphalt mix and about 12.4 kilometers of conventional dense graded asphalt mix placed on an existing 2 lane, Forest Service road. Both mixtures were placed on an existing 100 mm open-graded emulsion mixture (OGEM) pavement. The crumb rubber mix was placed in nominal thicknesses of 45, 65, and 90 mm while the conventional mix overlay was 90 mm thick. Both 90 mm overlays were placed in two equal lifts. A control section approximately 0.8 km in length adjacent to the rubber section was used for comparison. The gradation and other mix information on the two mixes can be seen in Figure 1. Additional details of the project construction can be found in Reference [2].

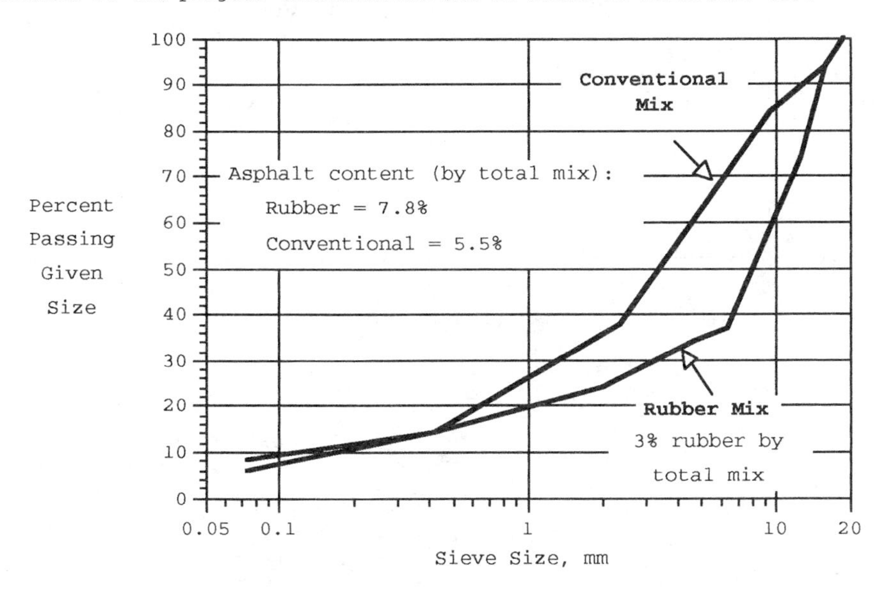

Fig. 1-- Aggregate gradation of conventional and crumb rubber modified asphalt mixtures used in Mt. St. Helens.

Testing

Considerable laboratory and field testing was conducted during the first 3 years following construction. Field evaluation of the sections included skid and texture testing and deflection and roughness measurements. A detailed discussion of the results can be found in Reference [3]; however, no significant difference was found between the crumb rubber and conventional mixtures during the first 3 years of testing. Laboratory diametral fatigue testing of cores taken from these sections showed that the rubber-modified materials had a higher fatigue life than the control mixture (Figure 2).

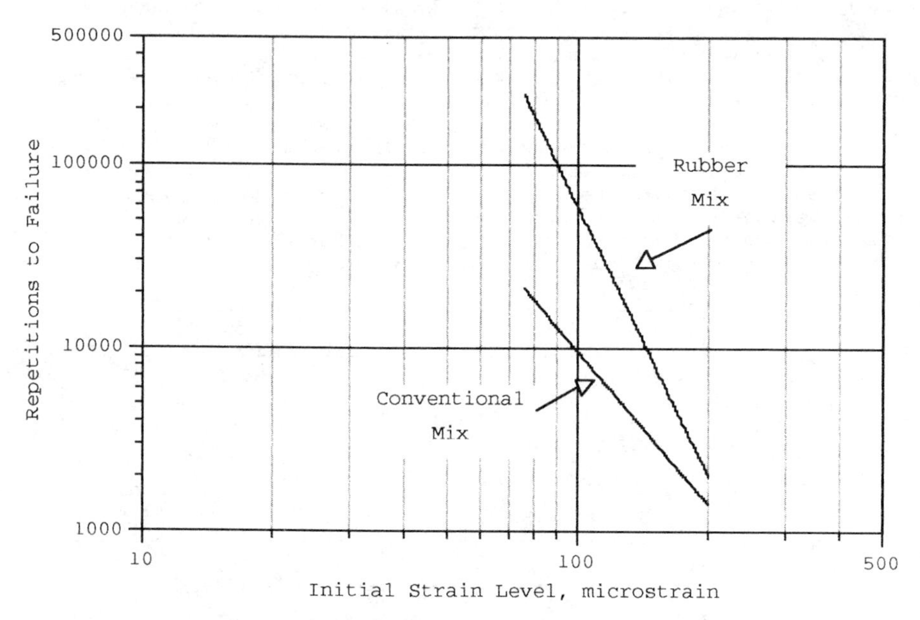

Fig. 2--Laboratory fatigue lives of conventional and crumb rubber modified asphalt mixtures used in Mt. St. Helens.

Cores were recently taken from the 45, 65, and 90 mm rubber sections in August, 1992. These cores were subjected to specific gravity measurement (ASTM D-2726) and repeated load diametral modulus and fatigue testing (ASTM D-4123). The results of the 1992 modulus testing and air void calculations are shown together with the earlier data in Table 1. The increase in modulus and decrease in air void percentage evident in the first three years appears to have stabilized during the last six years. Although fatigue tests were conducted on the 1992 cores, the results were highly variable. Furthermore, testing was conducted only at an initial strain of 200 microstrain precluding any meaningful comparisons to previous data.

TABLE 1--Modulus and air void data from rubber modified mixes.

Date	Number of Samples	Average Modulus, MPa	Air Voids, percent
September 1983	17	1550	4.9
November 1984	15	3280	3.6
July 1985	8	4030	2.4
June 1986	18	3050	2.8
September 1992	9	3300	2.3

Performance

Visual condition surveys were conducted yearly from 1983 to 1986 and again in 1992. Both the control and rubber mixture sections were in excellent condition with no evidence of rutting, alligator cracking, or other distress noted. Also, there were no discernible differences between the 45, 65, and 90 mm thick rubber-modified sections.

Traffic count and classification information were gathered for only the first 3.3 years following construction. If linearly extrapolated forward to the present (1992), then the total applied equivalent single axle loads (ESAL) in both directions is approximately 700,000. An estimate of the load carrying capacity of this pavement can be made using the AASHTO Design Guide [4]. These computations indicate the pavement with the thinnest overlay (45 mm) should carry more than 3.5 million ESAL before the present serviceability index (PSI) drops below 2.5. Obviously the pavement has been subjected to only a fraction of the loads which would cause a serviceability-based failure. At the present rate of loading about 40 years of traffic would be required to accumulated 3.5 million ESAL.

Summary

This crumb rubber modified project continues to perform very well after nearly ten years of service. There are several factors which may contribute to the excellent performance. First, the rubber mixture has more than two percent additional asphalt (by weight) than the conventional mixture. It is generally accepted that within limits, higher asphalt contents result in thicker asphalt films coating the aggregates. Thicker films in turn result in improved performance provided instability in the mix is avoided. For this project, it appears that modification of asphalt concrete using crumb rubber does not result in stability problems.

Second, the modified material has improved fatigue characteristics compared to the conventional mixture as a result of the thicker asphalt films. Laboratory testing of cores from this project have demonstrated this relationship. Field performance through 1992 shows equivalent performance for the materials and given the relatively low traffic loading rate it seems unlikely that either material will fail due to fatigue.

The modified mixture on this experimental project cost approximately 75 percent more than the conventional mixture. These costs would be expected to decrease as the material becomes used more routinely. The additional costs may be justified in some projects; however to date, both materials used on this project have provided equivalent performance for nearly ten years.

OREGON DOT

This project was constructed in 1985 to evaluate the performance and cost effectiveness of nine asphalt mixtures, containing different additives. Two of the mixtures contained ground rubber, one using a wet process and the other using a dry process [5, 6]. There were also two control sections, one with lime treated aggregate and one without.

The tests sections are located in Central Oregon in a region where pavements often show distress such as thermal cracking after a few years and/or stripping. The area has cold winters, hot summers, large daily temperature swings, and frequent freeze-thaw cycles. The total project length is about 6.4 km with the rubber and control sections about 0.8 km in length.

The existing pavement prior to overlay exhibited extensive load and thermal cracking. There were also areas which had raveled due to a loss of adhesion between the asphalt and aggregate. The entire section was overlaid with an Oregon Class C-mix (37.5 to 50 mm thick) with the exception of the rubber mix using the dry process. The gradation and other mix information of the rubber mixes and the control sections are given in Table 2. Additional details on the project can be found in References 5 and 6.

TABLE 2--Aggregate Gradation of C-Mix and Crumb Rubber Modified Mix (Dry Process) for Bend Test Road

Gradation	C-Mix (Control)	C-Mix Rubber Modified	Rubber Modified (Dry Process)
19 mm	100	100	...
16 mm	...	...	100
13 mm	99	99	89
9.5 mm	89	89	76
6 mm	66	66	38
No 10	32	32	31
No 30	...	...	19
No 40	14	14	17
No 200	5.8	5.8	8.9
% Asphalt (Weight of Total Mix)	6.9	8.0	8.0

Testing

Considerable laboratory and field testing was conducted during the first three to four years following construction. Field evaluation of the sections included friction testing, and deflection and roughness measurements [6]. After three years in service, there was no significant difference in the friction numbers between the crumb rubber

l conventional mixtures (all were between 56 to 62). Similarly, the
erage pavement roughness values as measured using the Mays meter were
. similar after four years in service. There were however,
fferences in surface deflection. One of the control sections
nsistently yielded higher deflections indicating a weaker base and/or
bgrade.

Laboratory diametral fatigue testing of cores taken from these
ctions showed that the rubber modified (dry process) test sections
re superior to that of the other mixtures. Typical results are given
low:

TABLE 3--Typical diametral fatigue data.

x Type	Load Applications to Failure	
	Initial* (Fall 1985)	After 3 Years (Fall 1988)
ntrol ith lime)	4,986	14,400
ntrol ithout lime)	7,094	8,700
umb Rubber ry process)	15,942	29,450
umb Rubber et process)	4,171	4,600

*Test performed at 200 microstrain and 23°C

res were recently taken from the four test sections in August 1992.
ese cores were subjected to specific gravity measurements (ASTM D-
26) and repeated load diametral and fatigue testing (ASTM D-4123).
e results of the 1992 modulus testing and air void calculations are
own together with earlier data in Table 4. Although fatigue tests
re conducted on the 1992 cores, the results were somewhat variable and
t repeated herein.

TABLE 4--Modulus and air void data from Bend Test Road

x Type	Property	Fall 1986	Fall 1988	Fall 1992
	No. of Samples	2	2	2
ntrol	Voids (%)	6.2	5.6	...
	Modulus MPa	2050	2630	1380
	No. of Samples	2	...	3
y Process	Voids (%)	4.4	4.0	...
	Modulus MPa	2210	2780	2395
	No. of Samples	2	2	3
t Process	Voids (%)	5.8	7.7	...
	Modulus MPa	1050	1715	1555

Performance

Visual condition surveys were conducted yearly through the Fall of 1989, and again in 1992. Both the crumb rubber sections were in good condition with limited amounts of thermal cracking. Little or no alligator cracking and some raveling were present in the pavement made with the mixture using the dry process. Figures 3 and 4 show the typical condition for the rubber sections. The control sections showed more thermal cracking; however, the performance in terms of load cracking and durability was comparable with the control sections. Note though in Figure 4, the thermal cracking proceeds through the control mix on the shoulder and stops when it reaches the rubber mix in the traveled lane.

Fig. 3--Photo of crumb rubber mix (dry process) in July 1992.

Fig. 4--Photo of crumb rubber mix (wet process) in July 1992.

Summary

The crumb rubber products continue to perform very well after 7 years of service. Both the dry and wet process have excellent resistance to thermal cracking as compared with the control sections. The mixture utilizing the dry process has experienced ravelling in areas likely due to quality control problems with the crumb rubber during construction.

Both crumb rubber products were considerably more expensive (75 to 100 percent), but this is due to the nature of experimental projects. These additional costs may be justified in some projects, but the advantages of crack resistance after seven years of service have not yet justified the added costs.

PORTLAND OREGON

The Portland Metropolitan Services District (METRO) constructed three experimental projects utilizing rubber-modified asphalt materials in 1991. These projects were initiated in response to a State Legislative mandate requiring the use of waste tire rubber in transportation projects. Two of the three overlay projects were constructed on low to medium volume streets using a dense graded aggregate with added crumb rubber in a dry process mixture, hereinafter termed RUMAC [7] The third project was constructed in cooperation with the Oregon Department of Transportation on Interstate 84 east of Portland. This project used both the dry (RUMAC) and the wet processes[8]. Only the third project will be described in significant detail.

The section of interstate chosen for this experiment consists of 15 to 28 cm of asphalt concrete on an aggregate base. Average daily traffic is about 13,000 (one way) with 16 percent trucks for a total one-way ESAL count of 1.1 million in 1991 [8]. The existing pavement had moderate to severe alligator cracking, rutting, ravelling, and potholing. After localized patching, a 50 mm overlay was placed using four overlay materials as shown in Figure 5.

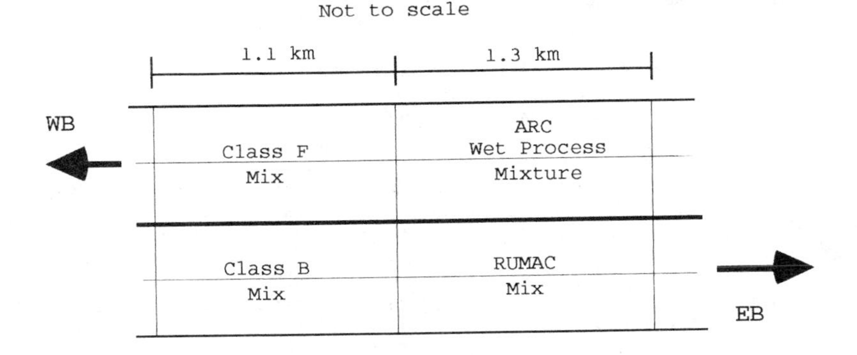

Fig. 5--Test section layout for Portland METRO experimental project [7].

The dry-process mixture (RUMAC) contained approximately 2 percent (by weight) rubber and 6.6 percent asphalt (by weight of mix)incorporated into a standard Oregon dense gradation [8]. Mixing and laydown of the RUMAC material proceeded without incident, however the contractor had difficulty achieving the specified minimum compaction of 94 percent. Variation in the rubber content were noted in some sampling sublots [7].

The wet-process mixture, termed ARC, was designed to provide a free-draining wearing course. As such, the ODOT open-graded (Class F) aggregate gradation was specified. Binder contents of between 8 and 11 percent were specified to insure adequate durability of the mixture. Final mix design binder content was set at 9.25 percent. The ARC material was mixed and placed without problem [7]. Compaction continued until the mixture reached a temperature of 80°C.

Cost comparisons are available for this project. As shown in Table 5, the cost of rubber modification of asphalt mixture is about 75 percent higher than conventional mixtures. These cost figures represent a single job which used fairly small quantities of the rubber-modified mixtures which would not be indicative of costs on larger jobs.

TABLE 5--Summary of material costs [7].

Mix Type	Quantity, metric ton	Price/metric ton	Price[a], $/m^2
RUMAC	1,510	$50.86	$6.10
ARC	1,510	$74.40	$7.49
Class F	1,790	$33.03	$3.70
Class B[b]	6,180	$32.23	$3.94

[a] Based on a 50 mm lift.

[b] Total quantity used, including level up course.

Laboratory Testing

Laboratory work on materials in this project included asphalt binder[8], diametral modulus, and fatigue testing. In addition, rutting potential tests were conducted using a Laboratoire Central Des Ponts et Chausses (LCPC) wheel track tester [9]. Original and residue binders were tested for penetration, viscosity and ductility. Additional tests were conducted as required by specification. Modulus, fatigue, and rutting test specimens were obtained from the RUMAC and Class B test sections.

The results of tests run on each binder used on the project are shown in Table 6. Additional tests results may be found in Reference [8]. The same performance-based asphalt (PBA-2) was used in the RUMAC and Class B mixtures [10]. A second binder type (PBA-5) was used in the Class F mixture[10]. Test results are also shown in Table 6 for the rubber-asphalt binder (ARC). Direct comparison of the test results is difficult since only 1 of the 11 tests were run on all binders and base asphalts were different for the three binders.

TABLE 6--Average binder test results [8].

Tests	PBA-2	PBA-5	ARC
Penetration @ 4°C, 100 g, 5s, residue, dmm	7	5	...
Penetration @ 4°C, 200 g, 60s, residue, dmm	25	22	28
Penetration @ 25°C, 100 g, 5s, residue, dmm	51	40	...
Absolute viscosity @ 60°C, original, poises	1490	2470	...
Absolute viscosity @ 60°C, 30 cm Hg vacuum, residue, poises	3940	7220	...
Absolute viscosity @ 60°C, 30 cm Hg vacuum, recovered, poises	3430	...	...
Absolute viscosity ratio, residue/original	2.7	3.0	...
Kinematic viscosity @ 135°C, original, centistokes	339	425	...
Kinematic viscosity @ 135°C, residue, centistokes	528	684	...
Ductility @ 7°C, 1 cm/min., residue, cm	25+	14	...
Ductility @ 25°C, 1 cm/min., residue, cm	100+	100+	...

Diametral resilient modulus and fatigue tests were conducted in accordance with ASTM D4123 with these variations, 1) test temperature of 23°C and 2) a single initial strain value of 200 microstrain [9]. The average of three cores are shown in Table 7. The modulus of the RUMAC mixture is lower than the Class B mix. This is to be expected since the RUMAC mix has a higher binder content and the presence of the rubber tends to soften the mix. Furthermore, the average air voids content of the rubber mixture was 10.2 percent compared to 6.9 percent for the Class B mix.

Fatigue testing demonstrates that the RUMAC mix has superior laboratory fatigue characteristics. However, the relationship between laboratory and in-service fatigue performance has not been established for rubber modified asphalt materials.

TABLE 7--Average test results on cores taken from RUMAC and Class B sections[9].

Material	Thickness, cm	Bulk specific gravity	Resilient modulus, MPa	Fatigue life, repetitions
RUMAC	4.6	2.240	870	4085
Class B	4.1	2.403	1000	1815

To evaluate the rutting potential of the mixes, two slabs were removed from both the RUMAC and Class B sections. After trimming the slabs were subjected to 20,000 passes of a 700 kPa tire loaded to 6.7 kN. The test temperature was 40°C. The final rut depths after 20,000 load applications were 9.2 and 8.4 mm for the RUMAC and Class B mixtures, respectively. These values are not statistically different at the 95 percent confidence level [9]. The development of the rutting is shown in Figure 6

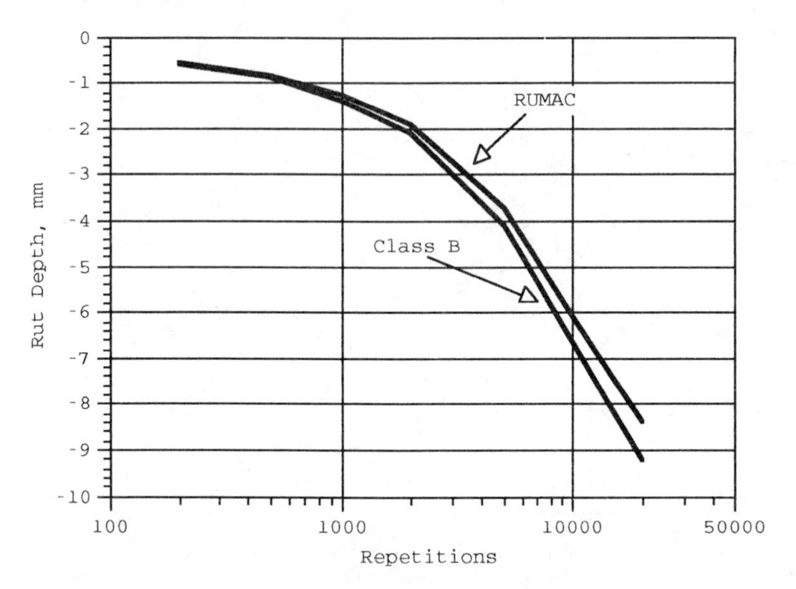

Fig. 6--Development of rutting in laboratory tests of RUMAC and Class B mixtures [9].

Performance

The experimental project was visually surveyed shortly after construction. All four test areas were reported to be in good condition with the exception of about 90 m of the RUMAC section [8]. This area showed significant ravelling. Although not conclusively determined, the problem was attributed to construction conditions. Possible causes including low mix delivery temperature, the presence of moisture on the existing pavement, and a reduced overlay thickness due to an overhead structure.

Surface friction and roughness measurements were taken in all sections. Test values were typical of new pavements constructed in Oregon except for a slightly high value of roughness in the ARC section [8]. The cause of this roughness has not been determined.

Summary

The construction of this project went smoothly without major problems. The contractor had some difficulty achieving sufficient compaction of the RUMAC mixture, however all other materials were placed without complication. Laboratory testing shows the RUMAC to have better fatigue characteristics than the Class B mix. Early monitoring shows all materials to be performing well except for localized ravelling in the RUMAC section as noted. Both the rubber modified mixtures cost substantially more than conventional materials used in Oregon. It is still too early to determine if the additional cost is justified based on improved performance.

DISCUSSION OF FINDINGS

Based on the results of these three case histories, plus the experiences of others in Oregon, the following conclusions can be made:

1) Rubber asphalt mixtures can be built successfully, but maintaining quality control is important. Insufficient rubber will result in instability problems while excessive rubber will result in raveling and/or potholing. The latter was observed on one of the projects reported herein.

2) Laboratory tests and the field observations indicate that the presence of rubber in asphalt mixtures improves crack resistance. However, the benefits may not offset the added costs.

3) The sections reported herein are still performing well, Therefore the failure mode of these rubber-asphalt mixes cannot be accurately assessed. Additional monitoring is required to establish the lives of the projects.

4) The experimental costs of rubber asphalt mixtures can be expected to be 50 to 100 percent in excess of normal paving costs. As agencies begin to use these materials more routinely, the costs would be expected to decrease substantially.

The above findings do not completely eliminate the concerns or barriers to using crumb rubber in paving applications. These barriers may be broadly classified as economic and non-economic. Generally, the economic barriers refer to the higher costs or limited return on the investment, whereas the non-economic barriers pertain to the environmental impact and technical factors such as the quality of the processes and products. Table 8 summarizes the most frequently cited barriers to waste tire utilization for crumb rubber modifier (CRM) technology in asphalt paving applications.

While the FHWA has estimated that paving costs for CRM technology are likely to stabilize at 20 to 30 percent more than conventional asphalt paving materials, there are several state highway agencies that do not share this optimistic projection. The greatest cost increase is a result of the fact that binder contents for asphalt-rubber mixes typically require about 20 percent more asphalt cement than do conventional mixes. In general, the use of asphalt rubber hot mixes will have higher costs due to the following:

- increased binder content;
- addition of crumb rubber;
- increased energy requirements due to elevated temperature and extending mixing time;
- additional plant personnel/equipment for handling and blending constituents; and

- additional personnel/equipment at the construction site.

Typically, cost increases may be apportioned as follows: materials, 50-80 percent; energy, 15-20 percent; additional labor and equipment, 3-5 percent; and royalty fees (where applicable), 15-20 percent.

TABLE 8--Barriers to CRM Utilization for Paving Applications.

ECONOMIC	• initial costs are about double the cost of conventional materials • insufficient life cycle cost data • capital cost for equipment modification
NON-ECONOMIC	• insufficient and/or conflicting long-term field performance data • lack of uniformity in specifications for rubberized asphalt • patented processes limit competition • potential leaching from tire chips where roadbeds are exposed to saturated conditions • potential environmental concern with visible emissions emanating from asphalt plants in which crumb rubber is being used

SUMMARY

The paper has presented the findings from three case histories (two to ten years in age) in which crumb rubber was used in the construction of asphalt concrete pavements. The results clearly indicate that successful projects can be constructed. However, the data to date do not conclusively demonstrate the long-term durability of the mixtures evaluated. The only benefit reported from these three case studies is improved thermal crack resistance.

REFERENCES

[1] Heitzman, Michael A., "State of the Practice - Design and Construction of Asphalt Paving Materials with Crumb Rubber Modifier," FHWA-SA-92-022, Federal Highway Administration, May 1992, 118 pp.

[2] Lundy, J.R., Hicks, R.G., and Richardson, E., "Evaluation of Rubberized Asphalt Surfacing Materials: Mt. St. Helens Study," Transportation Research Institute, Oregon State University, Research Report No. 84-3, June 1984.

[3] Lundy, J.R., Hicks, R.G., and Richardson, E., "Evaluation of Rubber-Modified Asphalt Performance-Mt. St. Helens Project," Proceedings, Association of Asphalt Paving Technologists, Volume 56, Reno, Nevada, 1987.

[4] "AASHTO Guide for Design of Pavement Structure," The American Association of State Highway and Transportation Officials, Washington, D.C., 1986.

[5] Hicks, R.G., Martin, K.L., Wilson, J.E., and Allen, D., "Evaluation of Asphalt Additives: Lava Butte Road - Fremont Highway Junction," Transportation Research Record 1115, Transportation Research Board, 1987, pp. 75-88.

[6] Miller, B., Scholl, L.G., "Evaluation of Asphalt Additives: Lava Butte to Fremont Highway Junction-Final Report," FHWA-OR-RD-90-02, Federal Highway Administration, October 1990, 102 pp.

[7] Takallou, H.B., "Mix Design Guidelines for Rubber Modified Asphalt Concrete," TAK Associates, Portland, Oregon, 1991.

[8] Zhou, H., Nodes, S.E., and Martin, K.L., "Rubber Modified Asphalt Concrete in Oregon," paper presented at Road Builder's Clinic, Spokane, Washington, March 1992.

[9] Lundy, J.R., Scholz, T.V., "Rubber Modified Asphalt Concrete Laboratory Test Results," Oregon State University, Corvallis, Oregon, 1991.

[10] "Standard Specifications for Highway Construction," State Highway Division, Oregon Department of Transportation, 1991.

Maqbool A. Khatri[1], D. Fred Martinez[1], Fouad M. Bayomy[2], J.A. Salter[3], and W.T. Tang[3]

PERFORMANCE EVALUATION OF ASPHALT MIXTURES WITH GASIFIER SLAG AS FINE AGGREGATE

REFERENCE: Khatri, M. A., Martinez, D. F., Bayomy, F. M., Salter, J. A., and Tang, W. T., **"Performance Evaluation of Asphalt Mixtures with Gasifier Slag as Fine Aggregate,"** Use of Waste Materials in Hot-Mix Asphalt, ASTM STP 1193, H. Fred Waller, Ed., American Society for Testing and Materials, Philadelphia, 1993.

ABSTRACT: This study investigates the chemical, physical, and mechanical properties of gasifier slag produced through the coal gasification process. The study focuses on the use of gasifier slag as a full or partial replacement for crushed fine aggregate and/or sand in hot-mix asphalt concrete. Performance of these newly developed slag-asphalt (Slagphalt) mixtures is evaluated in comparison with Control mixes, i.e., conventional mixes without slag. The low cost of slag makes it a viable alternative as a substitute for natural sand. The angularity and relatively narrow particle size distribution of the slag will be attractive in some regions.

KEYWORDS: gasifier slag, coal gasification, hot-mix asphalt concrete, performance, fine aggregate, mix design, environmental effects, chemical properties, physical properties, sensitivity, fatigue

---

INTRODUCTION

Slag materials are used in hot-mix asphalt concrete (HMAC) primarily to replace part of the conventional crushed aggregate and/or sand, depending on the gradation, and physical and chemical properties of the slag. The use of slag materials in HMAC provides an alternative, low cost aggregate source.

Existing coal burning processes traditionally have emitted sulfur dioxide fumes which contributed to the nation's acid rain problem. Coal gasification provides a clean and efficient process for converting coal into fuel gas. The process is based on a dry feed, entrained-bed, high pressure,

---

[1] Respectively, Research Engineer and Vice President, Research and Development Division, Southwestern Laboratories, 222 Cavalcade Street, Houston, Texas 77009.

[2] Associate Professor, Department of Civil Engineering, University of Idaho, Moscow, Idaho 83843.

[3] Respectively, Staff and Research Engineers, Shell Development Company, P.O. Box 1380, Houston, Texas 77251.

high temperature slagging design. This process can handle a wide variety of coals, ranging from bituminous to lignite, in an environmentally acceptable way and produces a high purity medium-BTU gas which is attractive for use in power generation. The burning of pulverized coal in this gasification process results in two types of by-product materials: fly slag and gasifier slag.

The scope of work for this study was divided into the following phases:

Phase 1. Literature Review
Phase 2. Chemical Analysis
Phase 3. Physical Analysis
Phase 4. Mixture Design
Phase 5. Fatigue Evaluation
Phase 6. Mix Design Sensitivity
Phase 7. Environmental Evaluation
Phase 8. Performance Prediction

A brief overview of each phase covering the objectives, materials used, testing program, and test results is given in the following paragraphs.

PHASE 1. LITERATURE REVIEW

This phase of the project was aimed at conducting a literature survey to gather information on the production and potential uses of gasifier slag as a paving material. With the commercialization of modern coal gasification processes, utilization of gasifier slag is becoming an important issue. There is not much existing literature addressing the application of gasifier slag as a paving material. However, the properties of gasifier slag are similar to those of boiler slag, for which a number of laboratory and field studies have been completed. In the following, the literature concerning the use of boiler slag, which is relevant to the application of gasifier slag, in the paving industry is briefly reviewed.

Production of Boiler Slag

Boiler slag is generated by two types of wet bottom slag boilers used in the utility industry. One variety burns pulverized coal, and the other, a cyclone boiler, burns crushed coal. Both of these boiler varieties have an orifice in the base which can be opened to permit molten ash to flow into a water-filled ash hopper. The molten ash quenches in the water, crystallizes, solidifies and forms angular, black, glassy particles, usually ranging from 1/4 to 1/2 inch in size. This wet bottom ash is known as boiler slag.

In a wet bottom boiler, the boiler slag amounts to 40 percent of the total ash produced while in a cyclone boiler the boiler slag amounts to 80 percent of the total ash produced. Overall, it is estimated that approximately 25 percent of all ash produced in electrical power plants is boiler slag (*1*).

Engineering Properties of Boiler Slag

Boiler slag possesses desirable engineering properties that have contributed to its application as a promising replacement aggregate for bituminous paving construction. Some important properties of boiler slag are summarized below.

Specific Gravity--The specific gravity of a boiler slag is generally a function of its chemical constituents. The specific gravity values are generally found to range from 2.5 to 2.7.

Grain Size Distribution--The range of grain-size distribution for boiler slag varies considerably. The particles range in size from fine sand to fine gravel. The gradation in general can be described as uniform.

Compaction Strength and Permeability Properties--Published data on the range of compaction, strength, and permeability properties to be expected for boiler slag is limited. The maximum and minimum relative densities for boiler slag are reported to range from 1458 to 1762 kg/m^3 (91 to 110 lb/ft^3) and from 1137 to 1410 kg/m^3 (71 to 88 lb/ft^3), respectively (2). The shear strength of boiler slag varies with the degree of compaction. The angle of internal friction for boiler slag in a loose condition can vary from 38 to 42.5 degrees and can average 41 degrees. The coefficient of permeability has been determined to vary from 0.3 to 0.9 mm/sec (0.012 to 0.035 in/sec), depending on the gradation of slag and the compactive effort used.

Chemical Constituents of Boiler Slag--The major constituents of boiler slag are silica, calcium, alumina, and iron oxide. There are also smaller quantities of magnesium oxide, sodium oxide, potassium oxide, sulfur trioxide and other compounds (*2*).

Experiments and Experiences with Boiler Slag

Some of the applications of boiler slag in bituminous pavement construction, as found in the literature, are briefly summarized below.

Bituminous Paving Mixtures--Published information on the use of boiler slag as a major component in paving mixtures is limited. The use of boiler slag in wearing mixtures is permitted in the specifications of several states including West Virginia, Indiana, and Ohio and it has been used in cities such as Tampa, Florida, and Columbus and Cincinnati, Ohio. Boiler slag possesses desirable skid resistant properties such as hardness and angularity. These properties have been utilized in deslicking applications and wearing course mixtures (*3*, *4*).

Stabilized Bases--Wet bottom boiler slag was used as an alternate base course aggregate in the repaving of light-duty, rural secondary roads in West Virginia in 1972. A one year post-construction report indicated the performance of the bases was encouraging. However, one year is too early for pavement distresses to develop (*3*).

Pozzolanic Base Courses--Boiler slag has been used in preparing both mixed in place as well as plant mixed pozzolanic base courses. These mixtures contained, on the average, 5 percent by mass of hydrated lime, 35 percent fly ash and 60 percent boiler slag (*5*). An experimental lime-fly ash-aggregate test section using boiler slag was placed on Illinois Route 9 in 1979. The test section was approximately 4 miles long and used a mix comprising 3 percent lime, 27.5 percent fly ash and 69.5 percent boiler slag (*5*).

Surface Treatments--Boiler slag has been used with good success in various types of surface treatments and seal coats. Spread rate comparisons of boiler slag surface treatment with conventional materials have confirmed that boiler slag provides better coverage per mile than limestone chips (*5*).

Ice Control--Eleven states, including Illinois and West Virginia have indicated that they have used boiler slag as an ice control material. One of the largest known uses of boiler slag for ice control purposes is in the Ohio River Valley where more than 250,000 tons of boiler slag are used during the winter in northwestern West Virginia, eastern Ohio, and southwestern Pennsylvania (*6*).

Other Uses--The high permeability of boiler slag and its resistance to degradation as well as freezing and thawing, have been applied in diverse non-bituminous paving applications which include: a) use of slag as a drainage and backfill material (*6*), b) use of slag as railroad ballast (*7*), etc.

PHASE 2. CHEMICAL ANALYSIS

The work during this phase was divided into:

1. Determination of chemical characteristics by X-ray diffraction and metal contents in the slag material.

2. Determination of bonding energy (Gibbs free energy).

3. Study of stripping characteristics by means of the Texas boiling test.

The results of X-ray diffraction indicated that the material is non-crystalline or glassy phase. The sulfur content in the slag was 1.06 percent and the iron content was about 886 ppm. Typical limestone aggregate would contain 0.18 percent sulfur and about 23 ppm of iron. This data indicated that the slag material tested had higher sulfur content relative to conventional limestone aggregates. This is not expected to be deleterious to the performance of the gasifier slag as an aggregate since the materials are held in a glassy matrix.

The spectral analysis, performed to determine the toxicity of the slag material in accordance with the Environmental Protection Agency (EPA) limits, indicated that the metals analyzed have contents much lower than the regulatory thresholds.

Results of the bonding energy experiments are reported in Table 1. The measurements were made on mixtures of asphalt (Exxon AC-20) with slag as well as limestone and gravel aggregates. The data available from the Strategic Highway Research Program's (SHRP) A-003B contract on Granite and Basalt aggregates are also included in Table 1 for reference and comparison. The results revealed that bonding energy between the slag and the asphalt binder was low. This suggested an expected low performance of asphalt-slag mixtures with respect to stripping. However, the actual performance evaluation did not support this concern.

TABLE 1--Slag bonding characteristics

Aggregate Type	Gibb's Free Energy (mcal/g-2hr)
Limestone	250-350
Granite	100-350
Slag	54-58
Basalt	600-800
Gravel	300-400

Note: 700's - high energy, 100's - low energy.

The stripping characteristics of the slag were evaluated using the Texas boiling test. Several mixtures of slag and Exxon AC-20 asphalt cement, both with and without modifiers, were evaluated. These mixtures included:

1. Mixture of slag and 4 percent asphalt cement, no other treatment
2. Mixture of slag plus 1 percent hydrated lime and 4 percent asphalt cement
3. Mixture of slag with 4 percent asphalt binder modified with 1 percent by mass of an antistripping agent, Kling Beta
4. Mixture of limestone screenings and 4 percent asphalt cement

Results of the stripping test are reported in Table 2. The numbers reported indicate a subjective rating of the degree of stripping after completion of the test.

TABLE 2--Results of Texas boiling test (TEX 530C)

Blend	Stripping, percent
Slag with 4% AC (no additives)	75
Slag with 4% AC (1% hydrated lime)	60
Slag with 4% AC (1% antistripping agent)	50
Limestone screenings with 4% AC (no additives)	50

PHASE 3. PHYSICAL ANALYSIS OF SLAG

This phase was designed to evaluate the engineering and physical properties of the slag. The tests performed during this phase and the corresponding results are reported in Table 3. The physical properties of typical fine crushed aggregate (limestone screenings) used in the Houston area are also reported for comparison purposes.

Following conclusions were drawn:

1. The slag material had a gradation close to that of the limestone screenings and common field sand used in HMAC mixes.

2. The physical properties of slag were found to be within the general requirements for fine aggregates. It had a higher water absorption as compared to that of the limestone screenings.

Based on the above conclusions, it was decided that the slag could serve as a possible replacement for limestone screenings.

PHASE 4. MIXTURE DESIGN

The objective of this phase was to determine the optimum slag content for hot-mix asphalt concrete (HMAC) mixtures. The Texas method of mix design (*8*) was employed to optimize the mix constituents including the slag material. The optimum binder content was defined as that which produced 96 percent of the theoretical maximum density after compaction.

Several different aggregate blends were used including a Control blend (i.e., no slag). The percentage of each type material used and the final gradations of the various blends are given in Table 4. The blends were designed to satisfy the Texas SDHPT Item 340 Type "D" gradation (*9*).

Results obtained from Phase 4 are presented in Table 5. The results indicated similar stability values for both slag and Control mixes. However, asphalt content for slag mixtures is slightly higher than that for the Control mix especially for Mix #3 where limestone screenings and field sand were replaced by slag (39 percent slag). This can be attributed to the relatively higher absorption of the slag.

TABLE 3--Physical properties of slag material

Sieve Analysis (ASTM C 117 and C 136)			
Sieve Size	% Passing Slag	% Passing Limestone Screenings	
3/8 inch (9.5 mm)	100.0	100.0	
No. 4	99.5	99.6	
No. 10	85.4	76.1	
No. 40	17.2	35.8	
No. 80	5.8	11.5	
No. 200	2.0	3.4	
Notes: 1. In the above, properties of limestone screenings are provided for reference only. 2. The following properties are for slag only.			
Specific Gravity (ASTM C 127 & C 128)			
Property	Ret. No. 10	No. 10- No. 80	Passing No. 80
Bulk Specific Gravity	2.551	2.644	-
Bulk Specific Gravity (SSD)	2.611	2.667	-
Apparent Specific Gravity	2.713	2.707	2.671
Absorption, percent	2.4	0.9	-
Sand Equivalent (ASTM D 2419)	98		
L.A. Abrasion (ASTM C 131)	Grading = D Percent Wear = 42.1		
Dry Rodded Unit Weight (ASTM C 29)	91.7 lb/ft^3 (1,471 kg/m^3)		
Moisture Content (ASTM C 566)	4.1 percent		

Based on the results of this phase, the following conclusions were drawn:

1. The mixtures produced by replacing the limestone screening in the Control mix with the gasifier slag provided viable alternatives.

2. The mixes with slag required somewhat higher asphalt content due to higher water absorption of slag as compared to that of the limestone screenings (water absorption = 0.3 to 0.5 percent).

TABLE 4--Details of the blends used during Phase 4

Mix No.	Control	1	2	3	4	New Texas Specifications Item 340-D
Constituents, percent						
Type D Limestone	40	40	40	40	48	-
Type F Limestone	21	21	21	21	12	-
Slag	-	15	24	39	30	-
Limestone Screenings (LSS)	15	-	-	-	-	-
Field Sand	24	24	15	*	10	-
Gradation						
Sieve Sizes	Percent Passing					
1/2 in (12.7 mm)	100.0	100.0	100.0	100.0	100.0	85 - 100
3/8 in (9.5 mm)	95.2	95.2	95.2	95.2	94.2	50 - 70
No.4	61.2	61.2	61.1	61.1	63.3	32 - 42
No.10	38.4	38.4	37.0	37.0	36.6	11 - 26
No.40	24.3	24.3	18.2	18.2	15.0	4 - 14
No.80	5.9	5.9	4.8	4.8	4.3	1 - 6
No.200	1.1	1.1	1.2	1.2	1.2	1 - 8

* The 15 percent sand in Mix #2 was replaced by slag; gradation same.

TABLE 5--Summary of mix design results for slag mixes

Mix No.	Optimum Asphalt Content (percent)	Hveem Stability
Control	5.4	39
1	5.8	38
2	5.8	40
3	6.2	38
4	5.8	43

PHASE 5. FATIGUE EVALUATION OF SLAGPHALT MIXTURES

The main objectives of Phase 5 were:

1. To investigate the fatigue behavior of gasifier slag modified HMAC mixtures (Slagphalt) at low temperatures as well as room temperature to determine the effect of temperature variation on the mixture performance in fatigue.

2. To develop the fatigue parameters required for performance evaluation using the beam fatigue test.

A total of 36 beam samples were tested during this phase. The various factors and their levels included in the test matrix were:

1.	Temperature	-	39 and 77°F (approx. 4 and 25°C)
2.	Stress level	-	High, medium, and low
3.	Slag content	-	0 and 24 percent (the Control and Mix #2 as described in Table 4)

An average optimum asphalt content of 5.5 percent was used in preparing all the specimens. The beam specimens were prepared using a kneading compactor. Air void levels ranged between 5 to 6 percent. The standard four point fatigue test was used to test all the beam specimens in controlled stress mode.

The bending stresses and strains determined were plotted against number of cycles to failure (N_f) and are shown in Figures 1 and 2, respectively. The following equations were used:

For bending stress - N_f relationship:

$$N_f = K_1 [1/S]^{c_1} \qquad (1)$$

For bending strain - N_f relationship:

$$N_f = K_2 [1/e]^{c_2} \qquad (2)$$

where,
N_f = Number of cycles to failure
S = Bending stress
e = Initial bending strain
K_1 , K_2 = Fatigue constants
c_1 , c_2 = Fatigue exponents

The bending stress - N_f relationship in Figure 1 revealed that the Control mix has somewhat longer fatigue life as compared to the Slagphalt mix. The data, however, have a narrow band and the points are scattered around the two lines. This leads to the conclusion that no significant differences would be expected in fatigue life between the two mixes. The data also showed that this is true at both 77 and 39°F (25 and 4°C). The fatigue life at 39°F (4°C) is higher than that at 77°F (25°C) as expected.

The significant difference observed between the two temperatures for the stress - N_f relationship (Figure 1) almost vanishes for the strain - N_f relationship (Figure 2) for the Slagphalt mix. The same is not true for the Control mix, however. This variation may be attributed to the aging effect since samples of Control mix were stored for a longer period of time before they were tested at 39°F (4°C).

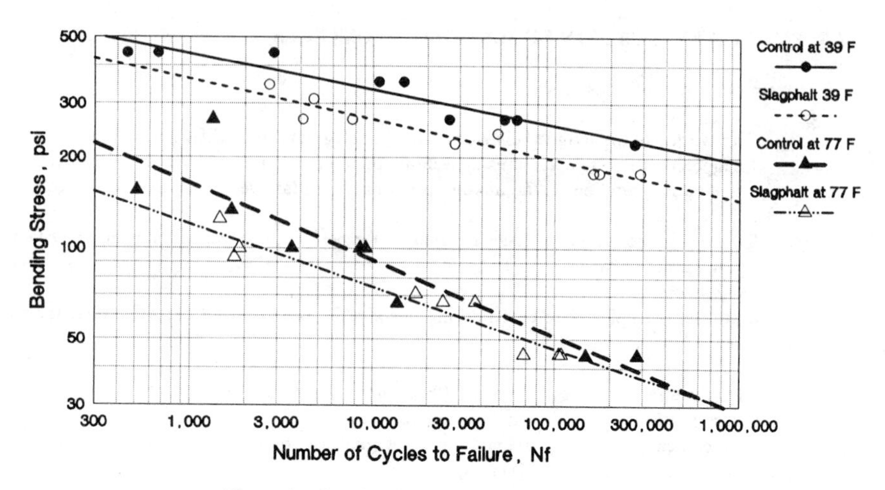

Figure 1 - Bending Stress - Nf Relationships

1 psi = 0.07031 kgf/sq cm
deg F = 1.8 * deg C + 32

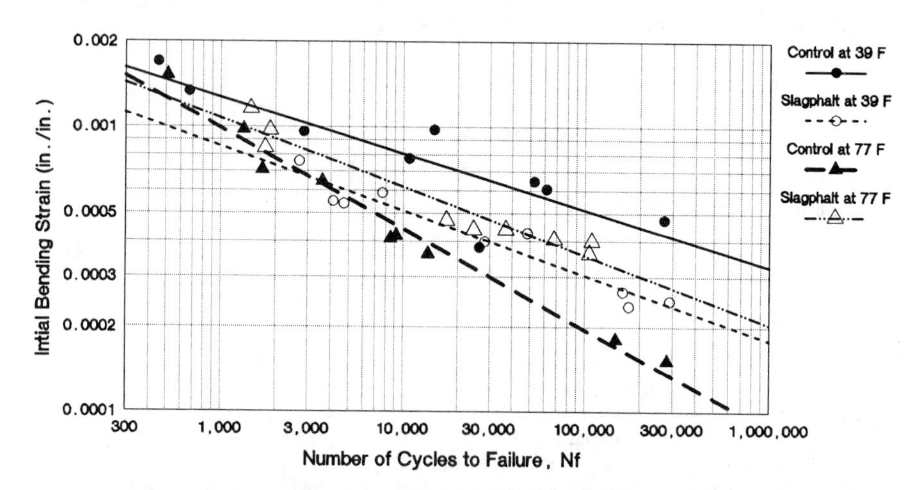

Figure 2 - Initial Strain-Nf Relationships

Based on the results of this phase, it can be concluded that, although there is some indication that the fatigue life of Control mixes is slightly higher than that of the Slagphalt mixes at 39.2°F (4°C), the difference at higher temperature is not significant.

PHASE 6. SLAGPHALT MIX DESIGN SENSITIVITY

The objectives for this phase of the project were:

1. To evaluate the influence of gasifier slag on paving mixtures.

2. To determine the effect of the slag on typical mixture properties.

3. To check if the various blends tested satisfied the mix design requirements.

4. To evaluate the economical impact of the use of gasifier slag.

A total of 34 mix designs were carried out during this phase. The various factors and their levels included in the test matrix were:

1. Aggregate blend - Blends 1 through 5
2. Slag content - None, Low, and High
3. Asphalt cement source - Exxon, Diamond Shamrock, and Coastal
4. Asphalt cement grade - AC-10 and AC-20

The composition of the various aggregate blends used during this phase is shown in Table 6. Blends 1 and 2 are both Type D wearing course mixtures and differ only with respect to their skid resistant characteristics. Blend 3 is a modified Type D mixture containing natural gravel aggregate. Blends 4 and 5 are 1 1/2 inch (38.1 mm) black base mixtures with the difference that Blend 4 contains limestone as coarse aggregate while Blend 5 contains natural gravel.

Fifteen 4 inch diameter (101.6 mm) x 2 1/2 inch (63.5 mm) cylindrical specimens were prepared for each mixture design. The mixtures were prepared in accordance with the Texas method of mix design. Optimum asphalt content was determined at 96 percent relative density. All mixtures were evaluated for Hveem stability (ASTM D 1560), resilient modulus (ASTM D 4123), and indirect tensile strength (ASTM D 4123).

Space does not permit presenting the data here, however, the major conclusions drawn from this phase can be listed as:

1. The use of gasifier slag seemed to be more effective in paving mixtures where it was substituted for natural sand as opposed to the manufactured sand. The use of slag also resulted in lowering the optimum asphalt contents for the black base mixes.

2. The asphalt cement source and grade did not have a significant effect on mixture properties with respect to a change in the slag content of the various mixtures.

3. Asphalt source was found to have some effect on the sensitivity of a mix as related to its slag content. The Exxon AC-20 was found to be better paired with the slag.

TABLE 6--Composition of aggregate blends used in Phase 6

Agg. Blend	1 1/2" (38.1 mm) LS	1 1/2" (38.1 mm) Gravel	SS	Gem Sand	Type D LS	Type F LS	LSS	Slag	Field Sand
B1-CTL	-	-	20	-	20	21	15	-	24
B1-G1	-	-	20	-	20	21	-	15	24
B1-G2	-	-	20	-	20	16	-	30	14
B2-CTL	-	-	-	-	40	21	15	-	24
B2-G1	-	-	-	-	40	21	-	15	24
B2-G2	-	-	-	-	40	16	-	30	14
B3-CTL	-	-	-	50	15	-	15	-	20
B3-G1	-	-	-	50	15	-	-	15	20
B3-G2	-	-	-	50	15	-	-	25	10
B4-CTL	35	-	-	35	-	-	-	-	30
B4-G1	35	-	-	35	-	-	-	10	20
B4-G2	35	-	-	35	-	-	-	15	15
B5-CTL	-	33	-	-	35	-	-	-	32
B5-G1	-	33	-	-	35	-	-	10	22
B5-G2	-	33	-	-	35	-	-	15	17

Notes: Bx - Blend x, CTL - Control, Gx - Gradation x

PHASE 7. ENVIRONMENTAL EVALUATION OF SLAGPHALT MIXES

The objective of this phase of the project was to conduct a limited study of the environmental effects of the Slagphalt mixtures. The work during this phase involved the use of Toxicity Characteristics Leaching Procedure (TCLP) to study the environmental effects of Slagphalt mixes.

Three samples of paving mix, two containing 24 percent gasifier slag (Blend #2, Table 4) and one Control (no slag), were tested using the TCLP procedure. All the metals as well as semi-volatile and volatile compounds were found to be far below the EPA limits.

PHASE 8. PERFORMANCE PREDICTION OF SLAGPHALT MIXES

The objective of this phase of the project was to develop hypothetical pavement designs for the Control and Slagphalt mixtures and to evaluate these designs to determine the effect of using gasifier slag on pavement life. The VESYS model (*10*) was used for predicting pavement performance.

The moduli and fatigue parameters were available from the results of Phase 4 and 5, respectively. The Poisson's ratio was assumed to be 0.35. Incremental static creep test was used during this phase to determine the creep compliance of the mixes (*10*).

Three design alternatives were adopted for each mix type resulting in a total of six design alternatives. The approach adopted was to assume traffic and soil conditions as well as base and sub-base conditions and calculate a pavement section to fit the assumed conditions. The AASHTO Design Guide (*11*) and the Asphalt Institute Design Manual (*12*) were used to develop a design for each mix. The evaluation of each design was performed using the AASHTO limiting strain criteria to determine fatigue and rutting life for each design.

The following input parameters were assumed:

- Traffic: 2 x 10^6 18-kip (8,182 kg) equivalent single axle loads (ESAL)
- Subgrade: soft clay with M_r = 7,000 psi (492 kgf/cm^2)
- Terminal serviceability: P_t = 2.5
- Analysis period: 20 years

Design alternative 1 was based on AASHTO Design Guide (*11*) using a reliability factor of 95 percent and an overall standard deviation of 0.35. Design alternative 2 and 3 were based on the Asphalt Institute Design Method (*12*). Design alternative 2 was a full depth pavement while in design alternative 3, a granular base was also introduced. The details of the various design alternatives are given in Figure 3.

The outputs generated by the VESYS program included damage and performance parameters. The damage parameters include the rut depth and the cracking area as a function of time. The rideability is represented by slope variance as a measure of pavement roughness. The performance parameter is represented by the present serviceability index (PSI).

Typical outputs for the different design alternatives are shown in Figures 4 through 15. Some of the major conclusions drawn from this phase of the project were:

1. For conventional flexible pavement systems (design alternative 1), the performance of Slagphalt mix was almost identical to that of the Control mix.

2. For full depth pavement systems (design alternative 2), rutting in the Slagphalt layer was higher by about 30 percent. The absolute values of the rut depth, however, were relatively smaller than those for conventional flexible pavements.

3. For the third design alternative, the performance of the two pavements was in between those for alternatives 1 and 2.

4. Slope variance, a measure of roughness, indicated that Slagphalt pavement might be rougher for the full depth pavement system.

5. Cracking did not vary significantly among the Control and Slagphalt pavements.

CONCLUSIONS

The following conclusions were drawn from the Slagphalt study:

1. A substantial amount of limestone screenings and field sand could be replaced with the gasifier slag making it an economically feasible alternative. However, it is expected to be more effective in paving mixtures where it was substituted for natural sand as opposed to the manufactured sand.

2. The Slagphalt mixes were found to be more cost effective as compared to the conventional HMAC mixes as they provide essentially equal performance for less cost (Slag is generally cheaper than limestone screenings by about $2-3 per ton.)

3. No environmental hazards are expected as a result of using gasifier slag in paving mixtures.

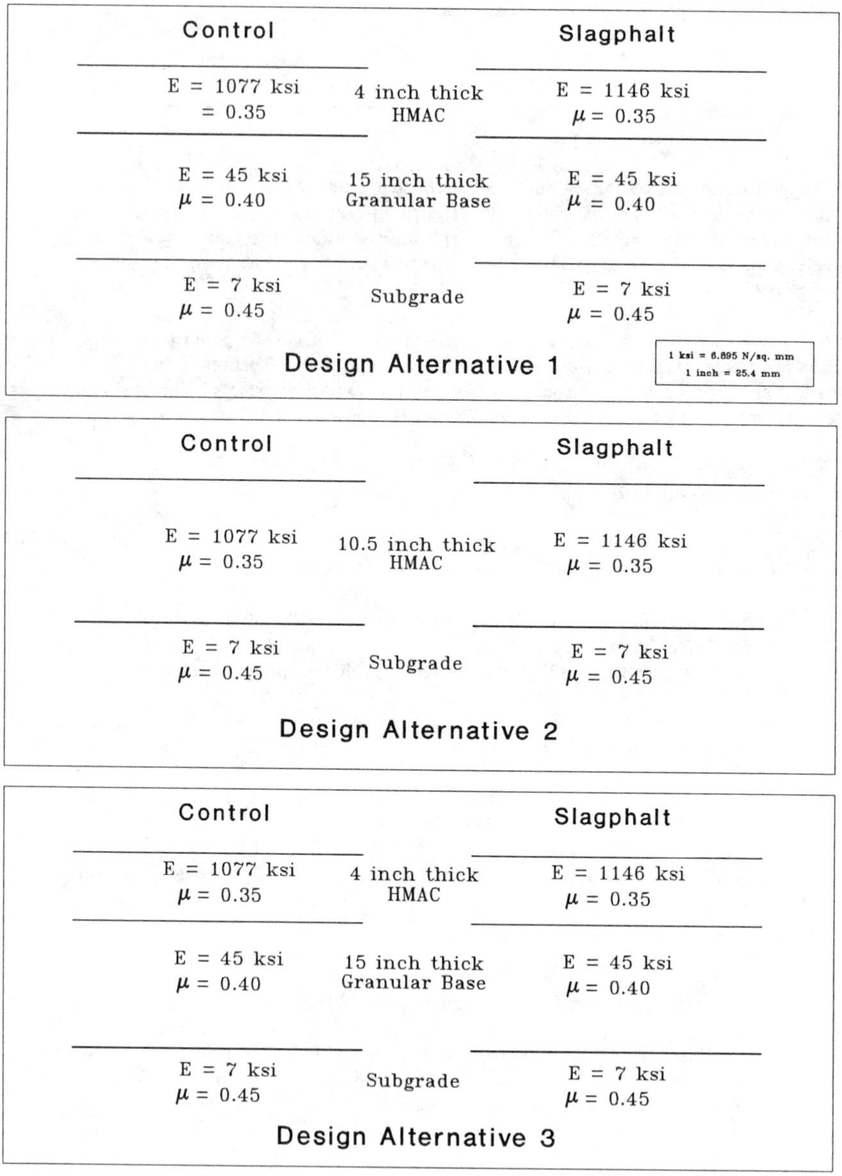

Control		Slagphalt
E = 1077 ksi = 0.35	4 inch thick HMAC	E = 1146 ksi μ = 0.35
E = 45 ksi μ = 0.40	15 inch thick Granular Base	E = 45 ksi μ = 0.40
E = 7 ksi μ = 0.45	Subgrade	E = 7 ksi μ = 0.45

Design Alternative 1

1 ksi = 6.895 N/sq. mm
1 inch = 25.4 mm

Control		Slagphalt
E = 1077 ksi μ = 0.35	10.5 inch thick HMAC	E = 1146 ksi μ = 0.35
E = 7 ksi μ = 0.45	Subgrade	E = 7 ksi μ = 0.45

Design Alternative 2

Control		Slagphalt
E = 1077 ksi μ = 0.35	4 inch thick HMAC	E = 1146 ksi μ = 0.35
E = 45 ksi μ = 0.40	15 inch thick Granular Base	E = 45 ksi μ = 0.40
E = 7 ksi μ = 0.45	Subgrade	E = 7 ksi μ = 0.45

Design Alternative 3

Figure 3 - Design Alternatives

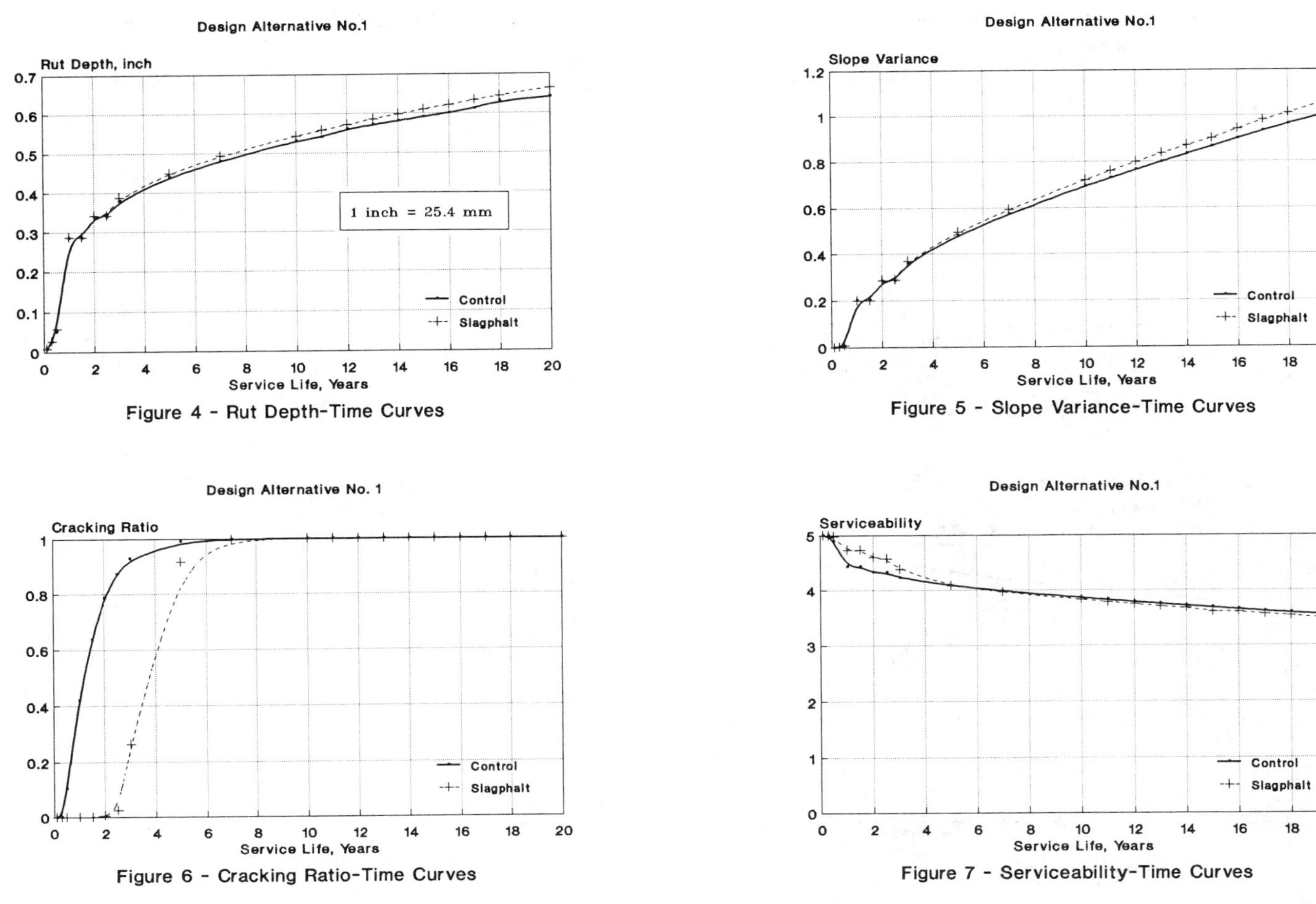

Figure 4 - Rut Depth-Time Curves

Figure 5 - Slope Variance-Time Curves

Figure 6 - Cracking Ratio-Time Curves

Figure 7 - Serviceability-Time Curves

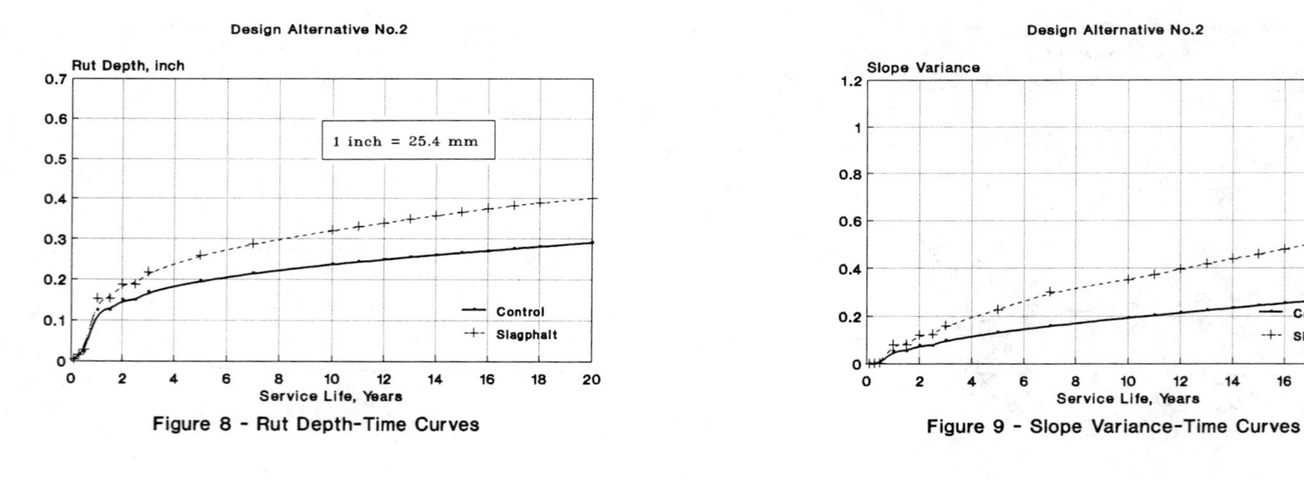

Figure 8 - Rut Depth-Time Curves

Figure 9 - Slope Variance-Time Curves

Figure 10 - Cracking Ratio-Time Curves

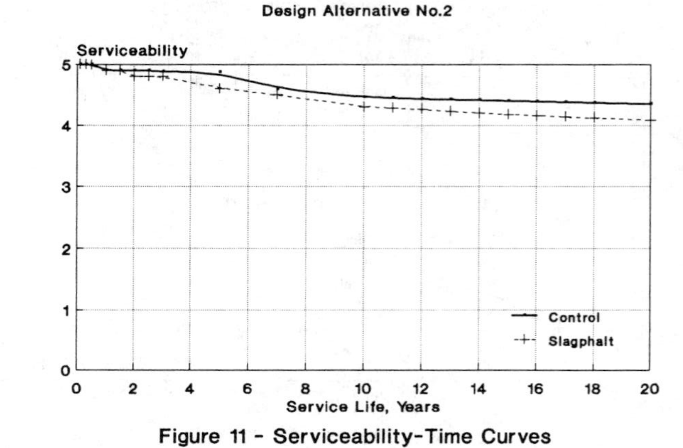

Figure 11 - Serviceability-Time Curves

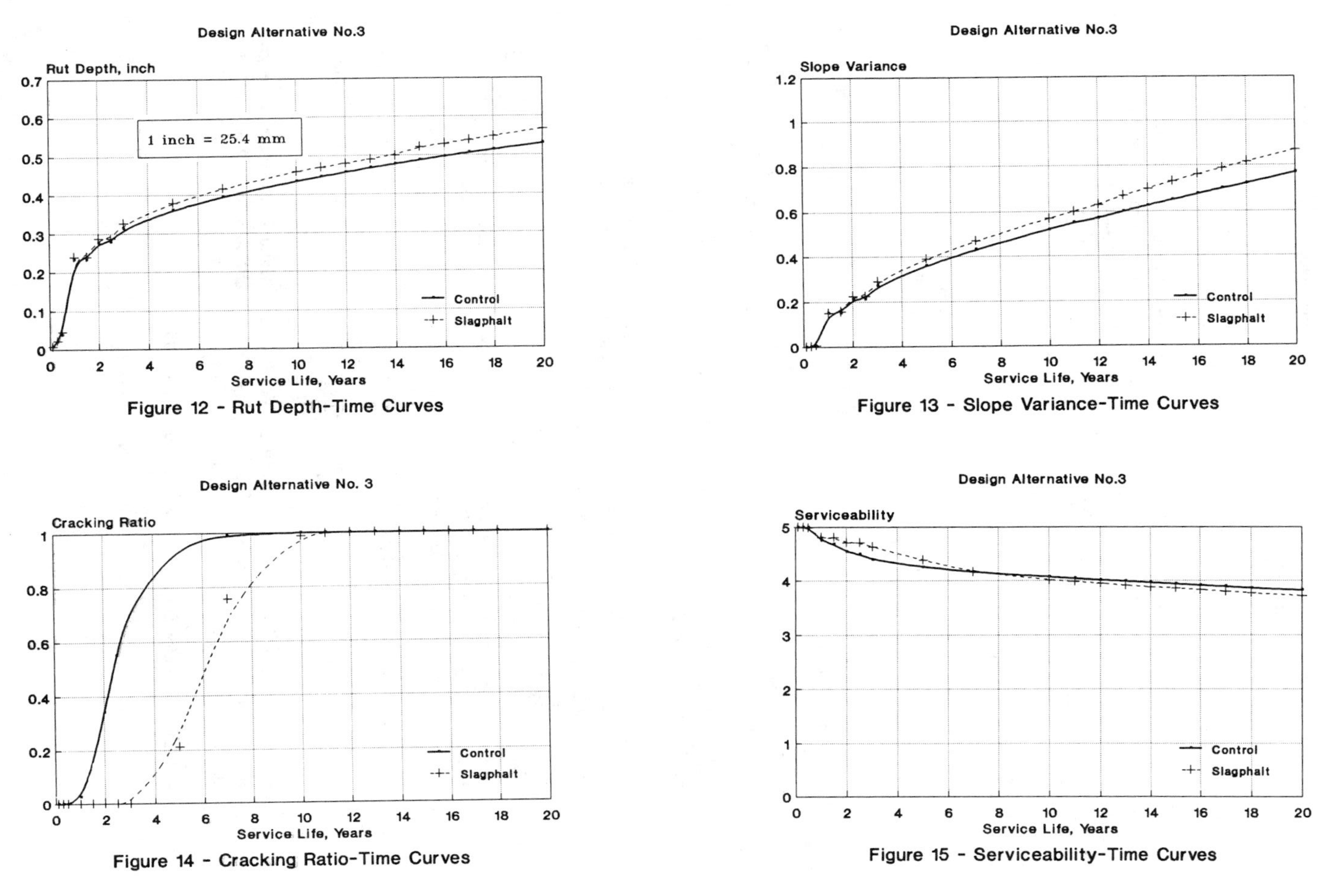

Figure 12 - Rut Depth-Time Curves

Figure 13 - Slope Variance-Time Curves

Figure 14 - Cracking Ratio-Time Curves

Figure 15 - Serviceability-Time Curves

4. The predicted performance for the Slagphalt mixtures was comparable to that of the conventional mixtures, i.e., containing no slag.

ACKNOWLEDGEMENTS

This study was conducted at the Research and Development Division, Southwestern Laboratories, Houston, Texas. Authors wish to acknowledge the Shell Development Company, Houston, Texas for their sponsorship and assistance in obtaining the gasifier slag for evaluation. Acknowledgements are also due to Dr. Xiaobo Pang of Southwestern Laboratories for his valuable assistance while performing the VESYS analysis.

REFERENCES

1. Miller, R.H., Collins, R.J., Ciesielski, S.K., Fergusson, W.B., Hill, V.J., Thomas, A.R., and Wallo, E.M., **"Waste Materials as Potential Replacements for Highway Aggregates,"** Valley Forge Laboratories, 1974.

2. Faber, J.H. and DiGioia, A.M., **"Use of Ash in Embankment Construction,"** Transportation Research Record, No. 593, 1976.

3. Moulton, L.K., Seals, R.K., and Anderson, D.A., **"Utilization of Ash from Coal-Burning Power Plants in Highway Construction,"** Highway Research Record, No. 430, 1973.

4. Brackett, C.E., **"Availability, Quality, and Present Utilization of Fly Ash,"** Proceedings of the First Ash Utilization Symposium, Pittsburgh, Pennsylvania, 1967.

5. Babcock, A.W., Collins, R.J., and Faber, J.H., **"Utilization of Power Plant Ash in Transportation Construction Applications,"** Proceedings of the Second International Ash Utilization Symposium and Exposition, Orlando, Florida, 1985.

6. Ash at Work, Vol. 2, No. 3, National Ash Association, 1970.

7. Cockrell, C.F., Shafer, H.E., and Leonard, J.W., **"News of Underdevelopment Methods for Producing and Utilizing Coal Ash,"** U.S. Department of Interior, Bureau of Mines, Information Circular No. 8488, 1970.

8. Texas State Department of Highways and Public Transportation Manual of Testing Procedures.

9. Texas State Department of Highways and Public Transportation Standard Specifications for Construction of Highways, Streets, and Bridges, 1982.

10. Kenis, W.J., **"Predictive Design Procedures, VESYS Users Manual - An Interim Design Method for Flexible Pavements Using the VESYS Structural Subsystem,"** Federal Highway Administration Report No. FHWA-RD-77-154, 1978.

11. AASHTO Guide for Design of Pavement Structures, 1986.

12. The Asphalt Institute Thickness Design Manual (MS-1).

Gene R. Morris[1]

TRUE COST EFFECTIVENESS OF ASPHALT-RUBBER PAVING SYSTEMS

REFERENCE: Morris, G. R., **"True Cost Effectiveness of Asphalt-Rubber Paving Systems,"** Use of Waste Materials in Hot-Mix Asphalt, ASTM STP 1193, H. Fred Waller, Ed., American Society for Testing and Materials, Philadelphia, 1993.

ABSTRACT: Asphalt-rubber is a material developed in the late 1960's and which today is being used on a worldwide basis in numerous types of pavement rehabilitation systems. The cost of the asphalt-rubber as a binder in either spray applications or hot mixed concrete is 3 to 4 times the cost of conventional asphalt cement. Because of these high material costs, many engineers believe these materials are only economical when life costs are considered. Further, some authors have written reports comparing costs of asphalt-rubber systems to conventional rehabilitation methods ignoring the basic applications of these materials. This paper presents various rehabilitation systems and case histories where asphalt-rubber has been shown to be economical on both first cost and life cost analysis.

KEYWORDS: Asphalt-rubber, pavement systems, asphalt-rubber concrete, seal coats, interlayers, SAM, SAMI, cost analysis.

BACKGROUND

The initial development of asphalt-rubber began in the mid 1960's when Charles McDonald, then City of Phoenix Materials Engineer, began searching for a method of maintaining pavements that were in a failure condition as a result of fatigue cracking. McDonald's early efforts resulted in the development of small, prefabricated asphalt-rubber patches that he called "Band-Aids". These patches were generally 24" x 24" (0.61m x 0.61m) and consisted of asphalt-rubber placed on paraffin coated paper with 3/8" (9.5mm) chips embedded.

Recognizing that fatigue cracking generally occurred in larger areas than small patches could handle, the concept

---

[1] Gene R. Morris, P.E., Technical Director for International Surfacing, Inc. 6751 West Galveston, Chandler, Arizona 85226.

was extended to full pavement sections by spreading the asphalt-rubber with slurry seal equipment followed by aggregate application with standard chip spreaders.[1] This process had two distinct construction problems. First, in order to achieve the desired reaction of the asphalt and crumb rubber in the limited time available in the slurry equipment, it was necessary to employ asphalt temperatures of 450°F (232°C) and higher. Second, the thickness of the membrane varied directly with the irregularity of the pavement surface. This resulted in excessive materials in areas such as wheel ruts and insufficient membrane thickness in between.

In 1968, technology had developed to the point that standard asphalt distributor trucks were employed to apply a uniform thickness of binder to the pavement. Although problems with distribution and segregation of materials were encountered on the early projects, these were recognized as primarily equipment limitations. Within the next few years equipment was developed with pumping, metering and agitation capabilities needed to handle the highly viscous asphalt-rubber materials.

Since these early developments, various uses of asphalt-rubber have been employed and several asphalt-rubber pavement systems have emerged. During the past twenty plus years, knowledge of asphalt-rubber material characteristics and quality control procedures have also progressed. Although there is still much to learn, technology and equipment have developed to the extent that asphalt-rubber can be designed and constructed with a level of confidence at least equivalent to standard paving procedures.[2]

This development period of over twenty years has not been smooth; it has included numerous failures and at least an equal number of highly successful projects. Most of the failures can be traced to insufficient technology, over-selling, lack of quality control, or resistance of the highway community to adapt use of unconventional materials.[3] The private companies involved in asphalt-rubber were relatively small and either did not recognize the need for development of technology or did not have the resources to finance such development. In more cases than one would like to acknowledge, the level of engineering competency applied to project design and construction has been woefully inadequate. In spite of these birthing pains, use of asphalt-rubber has spread worldwide. The technology has developed to provide an array of pavement systems that, when properly employed by the engineer, can result in significant initial and long term cost effectiveness. This paper will address the cost effectiveness of various asphalt-rubber paving systems.

Over the past several years, numerous authors have reported on the performance of asphalt-rubber, particularly the application of asphalt-rubber seal coats (Stress Absorbing Membrane, SAM) and asphalt-rubber interlayers (SAMI).[4] [5] [6] [7] [8] [9]

In the last two or three years, others have prepared reports analyzing cost benefits of asphalt-rubber systems based on the previously noted performance reports. Most of these reports have ignored some very basic considerations, thus their validity is highly questionable.[10] [11] [12]

SPRAY APPLICATIONS

The opening paragraph of this paper noted that Charles McDonald developed asphalt-rubber as a result of his search for a maintenance procedure to rectify fatigue cracking. The reason McDonald initiated this search was that no economical method of dealing with fatigue cracking existed at that time. The only viable methods available were either reconstruction or a thick overlay, neither of which were cost effective. The early experimental projects utilizing asphalt-rubber were all placed on pavements that were scheduled for reconstruction. The possible use of standard seal coats or thin overlays as a rehabilitation technique for these pavements had already been considered and rejected as a viable procedure.[13]

Typical of these early projects are the first three projects constructed by the Arizona Highway Department (ADOT). The first asphalt-rubber spray-applied by a distributor truck was placed on the ramps of the Black Canyon Freeway (I-17) in Phoenix, between Van Buren and McDowell in August of 1968. The pavement of these ramps exhibited severe fatigue cracking, considerable shrinkage cracking, and was beginning to pot hole. This location was chosen because reconstruction was already planned and budgeted and therefore, total failure of the experiment would not impose a significant problem. Although numerous problems occurred during construction, the performance of this SAM was extraordinary. It lasted 19 years with virtually no maintenance and was finally reconstructed in conjunction with freeway re-alignment.[14]

The second project constructed by the highway department was a six mile section of U.S. 60 - 70, approximately 90 miles northwest of Phoenix at Aguila. In 1972, this highway was carrying all of the traffic now carried by I-10. The pavement exhibited severe fatigue failure and although the pavement cross-section was still relatively uniform, pavement cracking level in this old, brittle mix was 100%. Again, this pavement was already scheduled for reconstruction. An asphalt-rubber chip seal was placed in June 1972. Subsequently, to improve rideability, a 1/2" (12.7mm) open-graded asphalt-concrete friction course was placed in 1976 (becoming a SAMI) and the pavement is still in service with minimal maintenance to date.[6]

The third project was constructed in 1973, on U.S. 89 north of Flagstaff, Arizona, between Townsend and Divide. In this case, reconstruction funds were unavailable and the District Engineer advised the State Engineer that he could not hold the pavement through the winter. Flagstaff is at 7,000′ (2,134m), elevation and experiences severe climate conditions with winter time temperatures down to -30°F. An asphalt-rubber chip seal was placed hoping to hold this pavement for eighteen months until reconstruction funds would become available. The project was finally rebuilt twelve years later in conjunction with realignment and widening to four lanes.[6]

These three projects all have one thing in common. Alternate rehabilitation techniques short of reconstruction or thick overlay were not even considered as viable alternatives. Yet every one of the authors has attempted to compare performance and cost effectiveness of asphalt-rubber constructed on totally failed pavements to chip seals and thin overlays constructed on sound pavements.

On all of these projects, the conclusions were that asphalt-rubber seal coats were effective at controlling fatigue cracking but were ineffective in controlling shrinkage cracking or low temperature cracking. It was observed however, that those cracks that reflected through were generally narrow, did not spall, and did not require extensive maintenance.[6] [9]

Since these early experiments, hundreds of asphalt-rubber chip seals have been placed to preserve totally failed pavements until such time as funds were available for reconstruction. Many of these pavements are still in service after twenty years with virtually no maintenance required.[15] It should be noted that most of these pavements were severely cracked because of extreme hardening (oxidation) of the asphalt concrete. Local areas of based failure were usually corrected. If one is to perform a realistic cost analysis of this system, the only comparison that can be made is of asphalt-rubber SAM versus new construction or at least the amortized cost of new construction. To compare the cost to a standard seal coat or a thin overlay, neither of which were viable alternative treatments, is not only wrong but in fact ludicrous. The cost of asphalt-rubber SAM is obviously less than reconstruction or a thick overlay.

Even if we consider those asphalt-rubber SAMs placed on roadways that exhibited some cracking but were not yet candidates for reconstruction, the various cost analyses presented are lacking. In Arizona, for example, the criteria for using a SAM or SAMI was that any pavement with more than 10% cracking, should be considered a candidate for a membrane.[13] If a standard chip seal was placed on a pavement with 10% cracking, its expected life with respect to reflective cracking would be at best one year. If reflective cracking is not the criterium for the need of a seal coat, then asphalt-rubber should not be used. However, if reflective cracking is the criterium then certainly the cost analysis should compare asphalt-rubber seal coats to the life of a standard seal coat placed on a badly cracked pavement.

Now, let's examine the other extreme of possible asphalt-rubber application - that of a SAM placed on new or sound, un-cracked pavement. In this case if only the life of the seal coat is evaluated then it is correct that the asphalt-rubber would have to exhibit twice the life or more of a standard seal coat in order to be cost effective. However, there is another extremely important consideration that should be included in the cost evaluation. Asphalt pavements crack and/or ravel because either the tensile strength or the tensile strain characteristics of the material have been exceeded. Rarely do new properly designed asphalt pavements crack. The failure generally occurs after the asphalt has aged to a level of stiffness such that a brittle failure occurs. If the asphalt in the pavement maintained its original material characteristics, then both fatigue and thermal types of failures could be greatly delayed. It has been shown in long term field performance that an asphalt-rubber membrane virtually eliminates aging of the pavement below. An analysis of 13" thick pavements with and without membranes at Sky Harbor Airport in Phoenix, showed that the pavement protected by the membrane experienced virtually zero oxidation while the viscosity of the asphalt in the unprotected pavement increased from 2,000 poises to over 200,000 poises.[15] Similar results were obtained by the Arizona D.O.T. from the analysis of an I-17 pavement south of Flagstaff, Arizona. If one is to properly evaluate the worthiness of an asphalt-rubber seal coat (SAM) as compared to a standard seal coat, then certainly this extension of life of the underlying pavement is a valid cost consideration. For example if the life of a 2 inch (51mm) thick pavement was extended for even 5 years then the initial increased cost of the asphalt rubber seal coat would be justified.

None of the authors have addressed this very real cost consideration in their analyses although some have made rather vague mention of it.

One other factor that has influenced the opinion of many state agencies on the performance of asphalt-rubber SAM & SAMIs was the use of these systems to control reflective cracking emanating from low temperature transverse cracks in the existing pavement. Perhaps these agencies were oversold by zealous salesmen. However, reports published by the Transportation Research Board as early as 1976, clearly[13] state that SAMs are not effective at controlling reflection of transverse cracks and that SAMIs are only somewhat effective. I firmly believe that systems can be developed to control reflection of low temperature cracks but it will take more extensive pavement preparation and multi-layered systems to accomplish this objective.

One specific application of SAMI is the three layer system developed for use as a thin overlay of portland cement concrete pavements. This system consists of a thin leveling course (1/2") (12.7mm) of hot mix followed by an asphalt-rubber SAM, and an open graded surface course. Total thickness of the system is 1 1/2" to 2" (38-51mm). The cost effectiveness of this system has been questioned by at least one author.[10]

There have been two major projects that utilized the three layer system. The first major project so constructed was on I-40 near Flagstaff, Arizona. The Flagstaff area is at 7,000 ft. (2,134 m) elevation and experiences temperatures as low as -30°F (-34°C). The section of I-40 involved was constructed in 1969, and by 1972, extensive problems had developed because of frost heave in the base and sub-base courses. The extent of distress was such that every panel in the outer truck lanes for four miles exhibited at least one crack. Obviously reconstruction and correction of the frost heave problem was the engineering answer. This answer, however, was totally unacceptable from a political standpoint. You simply don't rebuild 4 miles of interstate highway after only 3 years of service. Numerous possible rehabilitation treatments were considered but were rejected for various reasons. The solution that was chosen was to place a thin open-graded leveling course, an asphalt-rubber SAMI, and an open graded surface course. This concept was based upon preventing surface moisture from infiltrating the sub-base material, thus removing one of the necessary ingredients for frost heave to occur. This 3 layer system was placed in 1975, and provided low maintenance service until 1987, when delamination within the concrete itself required reconstruction.[16]

The second 3 layer system was placed on the Maricopa Freeway (I-10) in Phoenix in 1979. This PCCP pavement was constructed in the mid 1950's, and was a plain, undoweled pavement with joint spacing of 15'-17' (4.6m-5.2m). The joints in the pavement were not sealed and intrusion of incompressible materials had destroyed aggregate interlock with joint openings as wide as 1" (25mm) prevalent. Faulting of up to 3/8" (9.5mm) inch had occurred and some slab cracking existed. In addition, skid numbers had deteriorated to the point that the friction characteristics were dangerously low. Overall the service level of this pavement had deteriorated to the point where improvement in rideability and skid resistance was imperative.

Standard techniques for providing this improvement were: to grind and groove the pavement; construct a thick asphalt concrete overlay; or to reconstruct. Of these treatments, grind and groove would be the most economical. The aggregate used in the concrete is a hard, fine grained, quartzite and the cost of grinding was estimated at $8.00/sq. yd. ($9.57/m^2). In addition, this process would have required routing and sealing of all joints at a cost of approximately $2.00/ sq. yd. ($2.39/m^2). An experimental section employing the 3 layer system was placed in 1979, and subsequently in 1984, a four mile section of the freeway was rehabilitated at a cost less than $5.00/sq. yd. ($5.98/m^2). This section is in excellent condition to date. The only maintenance required has been minor crack sealing over those joints that were greater than 1/2" (12.7mm) in width. It should be noted that cost analysis at the time of construction indicated that this system would be cost effective if it provided slightly over 3 years of satisfactory performance.[17]

One other application of SAMI that has proven very cost effective is to provide a water proof membrane over expansive clay sub-grades. Brakey showed that the placement of a membrane directly over expansive clay sub-grades was very effective at reducing excessive volume change. This provided a design method for new construction, or complete reconstruction, but did not address the problem of existing highways constructed over highly expansive clays. The use of a SAMI, in conjunction with an overlay, provided a solution to the problem in existing highways. Again, the only alternate solution was total reconstruction.[18]

HOT-MIXES

To this point we have discussed only the spray applied asphalt-rubber systems. What about hot mixes? Are they also cost effective?

There are three types of hot mixes that are presently being employed - typical dense graded mixture, open-graded mixtures, and a relative new comer, the gap-graded or stone mastic mix. The first hot mix placed using an asphalt-rubber binder was a 3/4" (19.0mm) thick open-graded mix placed in April 1975, on St. Rt. 87 about 20 (32 km) miles north of Phoenix. The binder content was 10 1/2% by weight of the total mix. This section served maintenance free until 1991, when the 1 mile (1.6 km) section was included as part of a larger overlay project.

Since this first project in 1975, dozens of open-graded hot mixes with asphalt-rubber binder have been constructed and are performing extremely well. One of the most noteworthy projects is on I-19 south of Tucson where a 1" (12.5mm) overlay was constructed over plain, jointed PCCP in August 1988. To date this 1-inch (12.5mm) overlay has experienced zero reflective cracking and no rutting. Of particular interest is that noise measurements conducted by an independent laboratory showed average noise level reductions of 6.7 decibels, equivalent to a noise reduction of 78%.[19]

Again one must ask what alternate solutions were available to restore rideability and skid resistance to this portland cement concrete pavement at a cost of less than \$3.00/sq. yd. (\$3.59/m^2). The answer of course is none.

In 1983, Caltrans constructed an experimental project on Route 495 between Susanville and Ravensdale. This project included various asphalt-rubber mixtures, SAMs, SAMI, and control sections of various thickness. The performance of this project and several others have lead Caltrans to implement a policy that essentially decreases the thickness of an asphalt-rubber overlay to 1/2 that of conventional material.[20] Since asphalt-rubber hot mix is presently almost twice the cost of conventional hot mixtures the material cost of the structure is essentially equivalent. However, when you consider reduced aggregate quantities, construction time, traffic control and potential need to raise guardrails, signs, etc., because of overlay thickness, the asphalt-rubber will be lower in first cost. If you consider life cycle cost asphalt-rubber will virtually always be cost effective.

Another case history is the City of Phoenix where rehabilitation of the arterial streets consisted of fabric and 1 1/2 to 2" (38mm-51mm) of overlay. In 1989 the City bid as an alternate a one-inch (25mm) thick asphalt-rubber hot mix. This special mix was gap graded with 8% binder content by weight of total mix. These mixes have proved to be not only less in first cost, but have also outperformed the alternate system. The gap-graded mix developed by the City of Phoenix is the pavement for which Caltrans has determined that, from both structural and reflective cracking standpoint, overlay thicknesses can be reduced approximately 50%.[21] The economics of this asphalt-rubber hot mix were clearly demonstrated in a project near Kansas City, MO. Jackson County was responsible for maintenance and rehabilitation of the residential streets in the Salem East subdivision. The streets in 1989 were in a severely cracked condition and a project was scheduled to mill out 2" of the pavement and replace this with a 3" (51mm-76mm) overlay. The cost of this proposed design was estimated at $360,000 with a 60 day construction time. In lieu of this design, the streets were overlayed with a 1" (25mm) gap-graded asphalt-rubber hot mix with 9.5% binder. The total cost of the project was $170,000 and construction time consisted of 2 days - not 2 months. To provide the County assurance of performance, a 3 year bonded guarantee was required in addition to the normal 1 year construction bond. To date, these streets have required zero maintenance.

In summary, pavements are rehabilitated because of some specific problem. If pavement rehabilitation techniques are considered that truly provide a solution to the problem, then the use of asphalt-rubber systems can provide initial cost savings.

Those authors who are looking only at material cost comparisons should begin looking at the comparable cost of alternate pavement rehabilitation strategies. If systems are compared that truly provide solutions to the problems, then the use of asphalt-rubber can provide a first cost effectiveness in a majority of the cases. Asphalt-rubber will provide life cycle cost effectiveness virtually 100 percent of the time.

REFERENCES

[1] McDonald, C.H., "Recollections of Early Asphalt-Rubber History," National Seminar on Asphalt-Rubber, October 1981.

[2] Epps, J.A. and Gallaway, B.M., "Workshop Summary," First Asphalt-Rubber User-Producer Workshop, May 1980.

[3] Shuler, T.S., Pavlovich, R.D., Epps, J.A., and Adams, C.K., "Investigation of Materials and Structural Properties of Asphalt-Rubber Paving Mixtures," Volume I - Technical Report, Report No. FHWA/RD-86/027, Texas Transportation Institute, September 1986.

[4] Way, G.B., "Prevention of Reflective Cracking, Minnetonka East (1979 Addendum Report)," Report No. 1979-GW1, Arizona Department of Transportation, August 1979.

[5] Shnormeier, R.H., "Fifteen Year Pavement Condition History of Asphalt Rubber Membranes," City of Phoenix.

[6] Gonsalves, G.F.D., "Evaluation of Road Surfaces Utilizing Asphalt-Rubber - 1978," Report No. 1979-GG3, Arizona Department of Transportation, November 1979.

[7] Renshaw, R.H., "A Review of the Road Performance of Bitumen-Rubber in South and Southern Africa," National Seminar on Asphalt Rubber, October 1989.

[8] Delano, E.B., "Performance of Asphalt-Rubber Stress Absorbing Membranes and Stress Absorbing Membrane Interlayers in California," National Seminar on Asphalt Rubber, October 1989.

[9] Schofield, L., "The History, Development, and Performance of Asphalt Rubber at ADOT," National Seminar on Asphalt-Rubber, October 1989.

[10] State of the Art Synthesis Report, "Use of Ground Rubber in Hot Mix Asphalt," M.W. Witczak. Prepared for the Maryland Department of Transportation.

[11] Use, Availability and Cost Effectiveness of Asphalt Rubber in Texas, Estakhri, Fernando, Button, Teetes, Texas Transportation Institute.

[12] Shook, J.F., Takallou, H. B., and Oshinski, E., "Evaluation of the Use of Rubber-Modified Asphalt Mixtures for Asphalt Pavements in New York State," New York State Department of Transportation, December 1989.

[13] Morris, G.R. and McDonald, C.H., "Asphalt-Rubber Stress Absorbing Membranes: Field Performance and State of the Art," TRB Record 595, 1976.

[14] Morris, G.R., "Asphalt-Rubber Membranes, Development, Use, Potential," presented to Conference of the Rubber Reclaimers Association, Cleveland, 1975.

[15] Cano, J.O. and Charania, E., "The Phoenix Experience Using Asphalt-Rubber," National Seminar on Asphalt-Rubber, October 1989.

[16] Sarsam, J.B. and Morris, G.R., "The Three Layer System on Arizona Highways--Development and Analysis," 21st Idaho Asphalt Conference, November 1982.

[17] McCullagh, F.R., "A Five Year Evaluation of Arizona's Three Layer System on the Durango Curve in Phoenix," AAPT 1985.

[18] Way, G.B., "Dewey-Yarber Wash, Membrane Encapsulated Sub-grade Experimental Project," FHWA Contract DOT-FH-15-186, Arizona Department of Transportation, August 1982.

[19] Morris, G., "Asphalt Rubber Overlay Systems," International Road and Bridge Maintenance/Rehabilitation Conference and Exposition, Atlanta, 1988.

[20] Caltrans - Design Guide for ARHM-GG, Dated February 29, 1992.

[21] Cano, J.O., Charania, E. and Wong, D.C., "Gap-Graded Asphalt-Rubber Mix in the City of Phoenix," 4th Annual Conference and Road Show, San Antonio, TX, December 1990.

Author Index

Subject Index